NATHANAEL-ISRAEL ISRAEL, PhD

I0521000

From Science to Bible's Conclusions

OTHER BOOKS BY NATHANAEL-ISRAEL ISRAEL

Get them at your local bookstore, or online (e.g. on Amazon, Science180.com/books)

Turbulent Origin of the Universe
There is Only One Scientific, Simple, Safe, Trustworthy, Unexpensive, Brave, Practical, Nonconformist, Universal, Verifiable Formula that Accurately Decodes the Universe Formation … But You Are Not Using It

Reconciling Science and Creation Accurately
What Science Accurately Teaches about Creation and God's Existence that Atheists, Freethinkers, and even Most Christians Ignore … And How to Demonstrate it Without Taking Sides Between Rationality and Faith

Turbulent Origin of Chemical Particles
Why You Don't Have to Embrace Evolution, Big Bang, or Deny God to Scientifically Prove the Formation of All Chemical Particles

Origin of the Spiritual World
Top Secrets about the Origin of Everything in the Universe that Some Elites Have Hidden from You for Thousands of Years

Turbulent Origin of Life
Why You Don't Have to Embrace Evolutionism or Check Your Brain at the Door in the Name of Faith or Science to Accurately Decrypt the Origin of Life Using the Historic Formula of the Universe Formation

How God Created Baby Universe
What Children Must Scientifically Learn Early about the Universe Formation to Avoid Dangerously Abandoning God Later in Life Just Like Most College Students Who Embrace Evolution and Big Bang That Deny Biblical Creation

How Baby Universe Was Born
How to Scientifically Talk to Children about the Universe Formation and They will Know Forever How to Correctly Test the Intersection of Science and Faith

Science180 Accurate Scientific Proof of God
Can We Scientifically Explain the Formation of the Universe Through Natural Processes Without Evoking Evolution and Big Bang?

Mathematical Proof of God's Existence at the Intersection of Science and Faith.
The Scientifically Verifiable Cosmological Theory that Challenges the Big Bang Theory at the Crossroads of Reason and Religion THEY Want You to Ignore

More books written by Nathanael-Israel Israel can be found at Israel120.com/books

NATHANAEL-ISRAEL ISRAEL, PhD

Founder of Science180 and of Science180 University
Father of Science180 Cosmology and of Science180 Creationism

From Science to Bible's Conclusions

How Decoding the Universe-Origin by Properly Revisiting Scientific Data—That Top Scientists Collected but Wrongly Analyzed—Bizarrely led to the 3500 Years Old Biblical Account of Creation

Science180

Augusta
United States of America
www.Science180Publishing.com

Copyright © 2025 by Nathanael-Israel Israel
Visit the author's website at Israel120.com

From Science to Bible's Conclusions
How Decoding the Universe-Origin by Properly Revisiting Scientific Data—That Top
Scientists Collected but Wrongly Analyzed—Bizarrely led to the 3500 Years Old Biblical
Account of Creation

First edition: October 2025

Published by Science180
Augusta, Georgia (USA)
www.Science180Publishing.com

Book Cover and Illustrations by Nathanael-Israel Israel

ISBN: 979-8-9932150-5-1

Library of Congress Control Number: 2025920905

More books by the same author can be found at Israel120.com and Science180.com

For information about special discounts available for bulk purchases, please visit
Science180.com/discount for more details.

Science180 can bring authors including Dr. Nathanael-Israel Israel to your live or recorded
events. For more information or to book an event, please visit Science180.com/speaking

For any questions, please visit Science180.com/contact

To publish your book(s) with Science180 Publishing, go to Science180Publishing.com

To interview the author of this book, visit Israel120.com/interview
To donate, please visit Israel120.com/donate or Science180.com/donate.

Printed in the United States of America.

CONTENT

CHAPTER 1

WHAT DID THE MOST INFLUENTIAL NAMES IN PHYSICS MISS THAT THIS SCIENTIST GOT RIGHT ON HIS WAY TO ACCURATELY BREAK THE CODE OF THE UNIVERSE-ORIGIN?

You deserve to understand the origin of the universe, and by reading this amazing book, you will. Based on the most comprehensive theory ever developed for anyone seeking to crack the true origin of the universe, this elegant book will help you easily and quickly discover the process of the universe's formation from a fresh, internationally acclaimed angle.

1.1. Why I wrote this amazing book

- Why is it so challenging for people today to choose the right explanation of the origin of the universe?
- Must scientists and religious people stop trying to convince their listeners that they alone are always right and can help them make the right choice about the scientific explanation for the origin of the universe?
- How can focusing on where the scientific data faithfully leads us prevent the truth about the universe's origin from ever coming to light, even when scientists can collect more accurate data and use more advanced technology?
- Is there a scientific way to talk to evolutionists, Big Bang proponents, atheists, and all other freethinkers about the formation of the universe, and they will gladly beg to know more about the literal interpretation of the Biblical account of creation and God?
- How can freethinkers who rely on following reason and science know for sure that they are properly doing their rational homework concerning God?

- Why does the secularist world not care much if Christians and their leaders believe in Evolutionism, but they actually care much if they don't believe in the billions of years process?
- How can great scientists daring to challenge science, while faith leaders courageously questioning some religious doctrines, help humankind to holistically understand the origin of the universe?
- Why are opposing science and faith, choosing science only while rejecting or hiding faith, choosing faith only while avoiding science, religious people avoiding scientific criticism, scientists avoiding religious revelations, theorists wanting to collect more data before designing the correct cosmological models, all wrong ways to ever accurately understand the origin of the universe?

You may be tempted to answer these questions quickly yourself, but avoid ending up on the wrong paths that have cost some people their lives. It is better to get the accurate answer from the know-how expert, the author of many acclaimed books on the origin of the universe and its content, the standout expert who accurately decoded the universe's origin, including the scientific formula that forces science to meet the Bible. But where did they meet, and how can that intersection be properly demonstrated using unbiased science?

Indeed, questions about the origin of the universe have long baffled humankind. Therefore, throughout history, people of every age, nation, religion, culture, and background have pondered our origins, and many books devoted to the subject have offered various theories, anecdotes, myths, and narratives. Hence, understanding the true origin of the universe can affect your life in many ways. This book will empower you and make a real difference in your life.

When it comes to the origin of the universe, two views seem to dominate the others: (1) the Big Bang theory and (2) Creation by God. For instance, while proponents of the Big Bang and evolutionism theories (which are dominant in the academia) believe that the process of the formation of the universe took billions of years, some creationists (believers in creationism) support that the universe was formed within six days as mentioned in the Bible; whether the creation days were 24-hours each or not is a topic I will address later. During my investigation, I came across myths about creation, which are apparently wrong, but which, when properly compared with the holistic story of the formation of the universe, contain elements of truth about the beginning of the world, as if most ancient people have heard about a certain story regarding the formation of the universe which they changed as they wished, or which was changed along the generations.

After reviewing existing stories and theories about the origin of the universe found in the literature (some of which were told long before the scientific era), I found that the creation stories across the world can be classified into two categories: (1) the Judeo-Christian creation narratives (e.g., found in Judaism and Christianity) and (2) the non-Judeo-Christian creation narratives (e.g., found in Animism, Buddhism, Confucianism, Hinduism, Islam, and Evolutionism). I detailed the creation account of

2

Nathanael-Israel Israel: Known as the #1 International Authority that Truly Unlocked the Secrets of the Turbulence that Shaped the Universe

each of these religions in *"Turbulent Origin of the Universe"*, the detailed scientific version of this book. For instance, although Islam also supports a form of creationism, the Islamic account of creation differs from the Judeo-Christian narrative, which is based on the Book of Genesis in the Hebrew Bible (Old Testament) of the Christian Bible. The first chapter of the Bible's Book of Genesis summarizes most of what is known about the creation of the universe and its contents.

Among the Jews, Christians, and Messianic believers, the interpretation of the creation narrative varies and has given rise to many theories of creation, which can be classified into two groups:

- Old Earth creationism, which claims that the Earth is billions of years old, and
- Young Earth creationism supports that the Earth is *"between about 6000 and 10,000 years"* old.

The scientific community has extensively studied the origin of the universe. Although scientists have collected extensive data on celestial bodies, no one has yet provided an explanation that adequately accounts for the bulk of the scientific data and settles the disagreement between science and the creation narratives. For example, when those who believe in the billions-of-years process presented their theory, they made many assumptions that are hard for some people to accept. Likewise, although the creation narrative in the Bible said that God created the Earth on the 3rd day, and the Moon and the Sun on the 4th day, no human being, not even the Bible believers, has ever engaged in any scientific demonstration to prove that timeline. Most believers think that science cannot demonstrate the Biblical account of creation.

It is easy and common for both nonbelievers and believers to engage in heated debates about the timelines in their opponents' accounts of the origin of the universe. Still, no one has ever written an impartial scientific account that settles those disagreements with undeniable proof. After considering the existing accounts of the universe's formation, I was satisfied with neither the interpretations of the scientific evidence related to the origin of the world nor those of the creation narratives (even among some believers). In other words, I realized that both the scientific data collected about the universe and the creation stories have been misinterpreted by people, and, because of human mindsets, wrong explanations have unfortunately been proposed and believed by many throughout the ages. The scientific data were not properly explained, and no one has adequately explained the origin of the universe using the extensive information available in the scientific literature, without ignoring the religious narratives of creation. Considering human efforts to understand our origin, I felt like a study of the origin of the universe would be incomplete if it ignores also scientifically addressing the creation perspectives found in the religions of the world, not to show that all religions are right, but to use the scientific evidence to pinpoint the wrong religious accounts. Therefore, I decided to investigate the matter myself.

During my search for answers about the universe, I found myself delving into a vast amount of scientific data on celestial bodies. In the end, I wrote many books (learn more at www.Science180.com). Indeed, because of the diversity of human perspectives (even within the same religion or ideology) about our origin, it was impossible for me to put all my thoughts into one book. To tailor my findings to readers' backgrounds, interests, and needs, I organized them into several books. For instance, before publishing the first edition of this book, I devoted 12 years (2013-2025) to researching and writing books on the origin of the universe in a scientific language that some nonscientists may not fully understand. For instance, in addition to my scientific book on the formation of the universe, I also wrote another book on the origin of chemical particles and their clustering into molecules, minerals, and rocks. In another book, I handled the religious aspects of the origin of the universe. Furthermore, I also wrote a book on the origin of life.

Because I cannot replicate every detail of these books in the current one, I will highlight only a few key points relevant to the public, hoping those interested in the details can consult the original works. I really want everyone to get something out of my books, in whatever way they want. However, considering the amount of information I exposed in those books, and considering the complexity and controversiality of some of the things that need to be explained about the origin of the universe, I felt like the general public may have a hard time grasping the full picture of the origin of the universe if I do not break down the origin concepts using a simple language, vocabulary, and style that most people can relate to without making too much scientific or religious efforts to comprehend them.

Therefore, I decided to write this book not by going over everything I already addressed in the scientific and philosophical versions, but by highlighting a few key facts that I think most people are interested in knowing about the origin and formation of the universe. Likewise, in this book, I will not delve into the history of my research on the origin of the universe, as I have already covered it in ample detail in the scientific and religious versions. In the same way, instead of writing this book as a demonstration of all the evidence related to the formation of the universe, I wrote it as a story as I saw fit, knowing that those interested in lengthy demonstrations can follow up with the above-mentioned detailed books. The brevity of the demonstrations I present in this book does not mean that the long proofs are nonexistent or that I lack anything to say; rather, it is a strategy to avoid making this public book too long and too filled with jargon that the public may not know or be interested in. Without this approach, I could have ended up writing another long book, therefore defeating my purpose of using this book as a simple summary of the complex facts that I already established in the other books that may not fit the background of the nonscientists. In other words, this book that you are reading now, which I call the public version, is intended for everyone, and does not contain high-level scientific demonstrations that can confuse the illiterate. My goal here is to help those who may not have gone very far in secular education but can read and are seeking answers to questions about the origin of the universe and everything it contains. Nevertheless, those who have the chance to read the scientific and religious versions of this book will still find useful information in this popular version, for I

4

Nathanael-Israel Israel: Known as the #1 International Authority that Truly Unlocked the Secrets of the Turbulence that Shaped the Universe

CHAPTER 1: WHY AND HOW I WROTE A PUBLIC VERSION OF MY BOOK ON THE UNIVERSE-ORIGIN

painted the story in a different way and distilled my discoveries about the formation of the universe into a format and language digestible for the general public. To summarize, this public version of my book on the origin of the universe is not a compilation of all the evidence that I presented in my other books, but a summary of key points that can satisfy the hunger and thirst of non-scientists who are eager to know tangible facts and proofs about the origin of the world.

This book removed all serious challenges to those eager to understand where the universe came from. In fact, most people want an accurate, simple, straightforward, and nonpartisan universe-origin book that is free from jargon and difficult concepts only "known" by some experts. With laypeople in mind who lack the scientific background to handle the complicated, intimidating formulas and equations used in most cosmological theories, this book targets ordinary people who are tired of dealing with complex and useless theories of the universe's origin or dogma. It breaks down the technical aspects of the universe's origin and formation into scientific terms that even nonscientists can easily understand. This book bypasses technical knowledge that restricts non-experts from accessing the origin-related truth contained in the massive scientific data, and gets to the bottom of scientifically locked origin-related secrets regardless of your background. As you read this book today, you will indulge your mind to discover the truths revealed! Don't stop reading!

I discovered that the real reason scientists have been struggling to accurately understand the formation of the universe is that they have spent centuries collecting expensive, complicated, and massive amounts of data, but learned very little, if anything, about how to step back unconventionally to properly analyze it to decode the universe. Consequently, people learned to collect all kinds of data everywhere to build models and imaginary concepts that betray their discernment, but they never learned to unlearn wrong theories, nor learned how to stop trashing great raw data hidden in theories they dislike or misunderstand, never knew where to find and how to properly combine the fundamental variables without which it is impossible to ever clear the way so their data can properly work for and precisely lead them to the real origin of the universe. How can people abandon the dangerous theories they think are correct because they don't know any better ones?

Lucky you, that is where I, Nathanael-Israel Israel, the founder of Science180 (Science180.com), came in to properly reanalyze and put these costly, underrated data under control to provide the accurate, simple solution people have been looking for throughout the ages, but have ignored. "From science to bible's conclusions" is known as the # 1 universe-origin masterpiece of all time … and it contains the most accurate scientific formula that stood and will stand the test of time and of mathematics.

In *"From Science to Bible's Conclusions"*, you will:

- Get a world-class explanation of the 4 fundamental variables without which it is unquestionably impossible to ever decode the universe-formation scientifically

Science180: Breaking Cosmological and Traditional Universe-Origin, Life-Origin, and Chemicals-Origin Nonsenses

- Save time and money, and enjoy a life filled with the wonderful peace that the accurate understanding of the universe-origin can create
- Discover the errors in the scientific theories and religious belief systems about the universe's formation that are putting you at risk, and learn how to take control over cosmological threats lurking at the edge of your rational mind, faith, disbelief, or doubt
- Unlock the accurate scientific formula to rationally test the existence of God in a historic way that uncompromisingly satisfies both believers and skeptics (*Science180.com*/public)
- Get all you need to become a knowledgeable person who will never again need anybody else to explain to you the origin of the universe, for you will fully understand and articulate it yourself, and rationally know whether science is really at war with religion
- Receive deep insights that even those who went to university for years were not able to decrypt by themselves, so you can equip yourself to eliminate all forms of scientific and religious universe-origin prejudices
- Discover whether the scientific data finally confirms that the formation of the Earth was completed on the 3rd day, while that of the Moon and the Sun was on the 4th day of creation, like the Bible says, or whether the data proves that it took billions of years to progressively form the universe
- Understand the celebrated scientific formula that rationally puts to rest all debates about the relationship between science, faith, and all theories about the universe's origin, so you can properly develop yourself, expand your network, and shape your future

Keep reading this scientifically verifiable, bestselling book to finally get the accurate, jaw-dropping answer that has been rationally shaking both believers, skeptics, and all freethinkers. Don't stop. By the way, I have had the honor of being recognized as the #1 expert in the origins of the universe, life, and chemicals. I am the author of "*Turbulent Origin of the Universe*", "*Reconciling Science and Creation Accurately*", "*Turbulent Origin of Chemical Particles*", "*Turbulent Origin of Life*", "*How Baby Universe Was Born*", and "*Science180 Accurate Scientific Proof of God*". Visit Israel120.com to learn more about how I help scientists and laypeople properly decode the origin and formation of the universe, life, and chemicals, so people can live more effectively, nonstop.

Throughout my writing, wherever you see "universe-origin", please know that I meant "origin of the universe" or "the origin of the universe". Likewise, wherever you see "life-origin", please understand that I meant "origin of life" or "the origin of life". In the same manner, wherever I mention "chemicals-origin", please know that I am referring to "origin of chemicals" or "the origin of chemicals".

1.2. Essential topics to explain the origin of the universe

Although a lot of data is collected about the bodies in the universe, only a handful of variables are relevant to explaining its formation in simple terms. Here, I will present

6

Nathanael-Israel Israel: Known as the #1 International Authority that Truly Unlocked the Secrets of the Turbulence that Shaped the Universe

key topics that must be addressed to explain the origin of the universe to the general public. Indeed, in the scientific community, when it comes to the origin of the universe, many theories involving complex terms that the public cannot understand are proposed. From a scientific perspective, a lot of technical data needs to be handled before "convincing" some scientists. In my book *"Turbulent Origin of the Universe"*, I detailed these data. However, to explain the origin of the universe to the public, I did not feel that all the facts I discussed in *"Turbulent Origin of the Universe"* and *"Reconciling Science and Creation Accurately"* needed to be addressed here. In other words, given the scope of this book, I decided not to delve into complex terms that will not help most readers. Most of the technicalities of the scientific data may not even be understood by the public. Instead, there are key questions and facts that the public may want to understand in a language free of difficult scientific jargon. Therefore, here I focused on a few topics to pin down a story that can appeal to the public, while also providing an overall background on the hot topics related to the origin of the world. Consequently, the things I mentioned in this chapter as needing to be proven are not the complete list of what I tackled in all my books on the origin of the universe, but they are what I think the public may relate to better without annoying them.

Regardless of where they come from and what they believe in, most people accept that things in the universe can be categorized into 2 groups:

(1) the visible or physical world and
(2) the invisible or spiritual world.

It is important to mention that because people do not have the same spectrum of seeing, hearing, sensing, tasting, feeling, or thinking, what some people may clearly see with their eyes, others may fail to see even with the most sophisticated microscopes, telescopes, or other advanced tools. In the same manner, what is called the spiritual world is not denuded of physical things, and vice versa. Many things perceived as only physical are truly spiritual. Many things that were perceived as scientifically invisible a few hundred years ago can be "clearly" seen today just by using a microscope or a telescope. In the same way, many things that can be easily perceived through faith or belief remain invisible to those who rely solely on physical means. Therefore, although some people think that the invisible or spiritual world is only for religious people, it is worth noting that the boundary between the visible and invisible worlds is not clear-cut. In other words, the demarcation of the physical and the spiritual is not straightforward. This also implies that what people call "physics" is not just physics, but a field filled with spiritual and nonphysical facts that even the most famous physicists will never understand or explain if they rely solely on traditional science. This can also explain why many facts cannot be scientifically proven without a spiritual insight. Hence, in addition to *"Turbulent Origin of the Universe"* in which I detailed the "physical" aspects of matter, I also wrote 3 other books which expound on the spiritual world:

- "Reconciling Science and Creation Accurately"
- "Origin of the Spiritual World"

- *"Science180 Accurate Scientific Proof of God"*

In this book you are reading now, I will focus on the physical world, knowing that those who want to learn more about the spiritual world can consult the books I wrote on it.

As of today, most scientists consider the physical world to range from subatomic particles (particles smaller than atoms) to the largest galaxy clusters. The physical things in the universe are hierarchically clustered into structures. For example, some subatomic particles are collected into atoms, which in turn can be clustered into molecules and chemical compounds, which in turn can be clustered into minerals and rocks, which are constituents of celestial bodies. Examples of celestial bodies are planets, asteroids, satellites, planetary systems, asteroid systems, stars, stellar systems, galaxies, and clusters of galaxies. For those who are not familiar with these terms, I would like to note that a planetary system consists of a planet orbiting one or more satellites. Likewise, some asteroids also have their own satellites, and an asteroid system is an asteroid orbited by a satellite. A stellar system consists of a star orbited by planets, planetary systems, asteroids, and asteroid systems. In their turn, stellar systems can be organized into galaxies, which can also be organized into galaxy clusters. The Solar System, for instance, is a stellar system belonging to the Milky Way Galaxy.

During my research on the origin of the universe, I focused on planets, asteroids, satellites, rings, atmospheres, and crusts of the celestial bodies in the Solar System. I also dealt with stars and galaxies. In order from the Sun, the planets in the Solar System are Mercury, Venus, Earth, Mars, Jupiter, Saturn, Uranus, Neptune, and Pluto. The definition of a planet has been a controversial topic lately. For instance, from its discovery until 2005, Pluto was considered a planet. But, after the discovery of other celestial bodies whose sizes are closer to that of Pluto, a lot of debates have taken place. As of today, Pluto is not considered a planet by some astronomers, but most people still consider it a planet. Nevertheless, in my writings, I considered Pluto as a planet. Some planets are called terrestrial, meaning they are similar in size or composition to the Earth, and have a hard crust (i.e., Mercury, Venus, Earth, Mars, and Pluto); others are giant gas (i.e., Jupiter and Saturn), while others are giant ice (i.e., Uranus and Neptune).

Although many asteroids have their own satellite(s), when I talk about satellites in this book, I mean natural satellites, or natural moons of planets, not artificial or man-made satellites. Human beings have launched countless manufactured satellites into space, and some are used to improve communication, the internet, intelligence, and other military and commercial applications. As far as satellites are concerned in this book, I will be dealing with the 210 discovered in the Solar System as of 2021:

- Moon (Earth's only satellite)
- 2 Martian satellites
- 79 Jovian satellites
- 82 Saturnian satellites

Nathanael-Israel Israel: Known as the #1 International Authority that Truly Unlocked the Secrets of the Turbulence that Shaped the Universe

- 27 Uranian satellites
- 14 Neptunian satellites
- 5 Plutonian satellites

Next, I sketched the orbit of the main celestial bodies in the Solar System (Fig. 1).

Fig. 1: Sketch of the planets and main belt asteroids in the Solar System

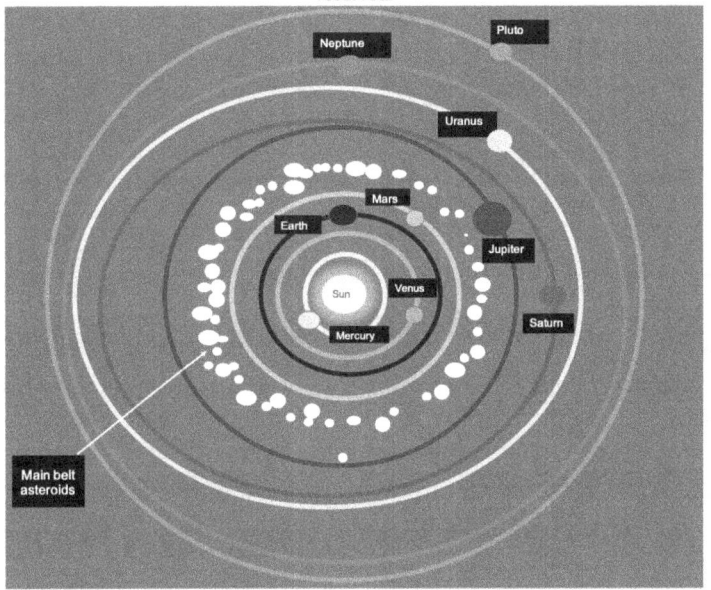

One thing is to want to explain the origin of the universe, but another thing is to know which elements or features must be investigated. For instance, if nonscientists talk about the origin of the universe, they would focus on the Sun, the Earth, the Moon, stars, planets, living things on Earth, and the night sky. But, scientifically speaking, there is more to explain in the universe than just what we see with the naked eye. For instance, throughout history, much data has been collected about celestial bodies that no human on Earth can see with the naked eye. A proper explanation of the universe's origin also needs to consider variables derived from those bodies. But given the narratives available about the universe's origin, I felt like most people do not even know the key questions to ask to properly explain the universe's beginning. To demonstrate the formation of the universe, I dealt with hundreds of variables, out of which I will focus on the following in this book:

- Axial tilt
- Density
- Eccentricity

- Escape velocity
- Gravity
- Kinetic energy
- Mass
- Orbital inclination
- Orbital speed
- Radius
- Rings
- Rotation period
- Semi-major axis
- Volume

If you do not understand the meaning of these variables, do not worry, for I will revisit them in the next chapters. Moreover, although many things can be said about each of these variables, to make this book as essential and digestible as possible, I removed details that do not add much to the story I think is relevant to the public.

Finally, in, around, and between the celestial bodies, chemical particles exist. Unlike celestial bodies, chemical particles are characterized using different variables. These chemical particles and celestial bodies interact with one another, and some of those interactions are mediated by what are termed fundamental forces, with gravity being the one the public knows best. Therefore, I will elaborate on these forces with an emphasis on gravity, and I delved into them in *"Turbulent Origin of Chemical Particles"*.

After reviewing key facts about celestial bodies in the universe, I will explain the underlying causes of their trends. Then, I will use my insights from those trends to argue for "Science180 Turbulence 1," my theory of the turbulent origin of the universe. Before concluding this book, I will compare the scientific evidence with the religious narratives and determine which creation story is correct. To learn more about other aspects of my research on the origin of the universe, please visit www.Science180.com, the website associated with this book and with others I have written.

CHAPTER 2

WHAT ARE THE 12 SIGNS THAT SOMEONE IS FINALLY READY TO PROPERLY DECODE THE UNIVERSE-ORIGIN?

You merit understanding the most important facts that can help decode the origin of the universe. In this chapter, I will present these facts in a format that you can fully comprehend. These facts are so important because the reasoning of most theories is based on postulates, which are propositions proven or assumed to be true and used as premises or starting points for arguments. Unlike some theories that are based on assumptions, my theory on the origin of the universe is based on established scientific facts verified over the years and accepted as axioms, meaning regarded as established and self-evidently true. I could have started telling a story about the beginning of the universe without first mentioning these facts, but I felt readers would appreciate my sharing the sources and evidence for some of the claims I will make in the following chapters. In other words, this chapter summarizes key scientific evidence underpinning the scientific demonstration presented in this book.

Many facts have been collected about celestial bodies in the universe, and if they were combined, many books could be written. Some of those facts contain no tangible information that the public can relate to, and even scientists struggle to tell a compelling story about them. Therefore, instead of compiling those facts here just to entertain the brain, I decided to focus on a few that can help me tell the story of the origin of the universe. Also, because of the limited space in this book and my desire to be concise, I will pay particular attention to only a few specifics, such as some important patterns in the universe, the distance between the celestial bodies, the speed, rotation, inclination, elongation of orbit, tilt of rotational axis, radius, and density of the celestial bodies. Although I had to reanalyze these variables, none of their raw data were collected by me; they were collected by spatial agencies such as NASA (the US government space agency) and other research institutions across the globe. After simplifying these variables, I will explain why their trends exist and what caused them. Then, I will logically link the scientific data and walk you through the thought process that led me to the conclusion I present in this book about the formation of the universe.

2.1. Organization of the universe as a set of primary and secondary bodies

Many patterns exist in nature, and one of them is the organization of celestial bodies as a nest of primary and secondary bodies. In a system of bodies, a secondary body orbits its primary body. For instance, in the Earth-Moon system, the Moon (the secondary body) orbits the Earth (the primary body). Likewise, in the Solar System, the planets and asteroids (secondary bodies) orbit the Sun (their primary body). Even on astronomical scales, scientific evidence shows that stellar systems orbit the center of their galaxy. Likewise, some galaxies are believed to orbit others. Moreover, in each system of bodies, the primary body is usually the largest body, while the secondary bodies are usually smaller. For instance, in the Solar System, the Sun is the largest body. Likewise, in the Solar System, the primary planets are always the largest bodies. But why is the largest body in each system of bodies located toward the center of the system, and why is it orbited by (smaller) secondary bodies? Later in this book, I will help you understand how these patterns relate to the formation of the universe.

2.2. Semi-major axis

Celestial bodies orbit their primary bodies at different distances, and most orbits are rarely a perfect circle. The semi-major axis is the average radius of the orbit of a celestial body. This means that the semi-major axis of a planet's orbit in the Solar System is the planet's average distance from the Sun. Similarly, the semi-major axis of a satellite is its average distance from its primary planet. But can the semi-major axis give a clue to the birthdate of the celestial bodies in the universe?

The semi-major axes of the celestial bodies I studied in the Solar System range from 9,378 km to 4.98 trillion km. The celestial bodies with the smallest semi-major axis are satellites of the smallest planets and satellites of the outer planets. To many Americans who prefer to express distance in miles, I would like to remind you that 1 mile = 1.61 kilometers. The semi-major axis of the Earth is 149,597,900 km; that distance is also called the Astronomical Unit (AU). In other words, the semi-major axis of the Earth is 1 AU. The semi-major axis of the planets in the Solar System varies between 0.39 AU and 39.5 AU (Fig. 2); the smallest value is recorded on Mercury (the closest planet to the Sun) and the highest one on Pluto (the outermost planet). This means that the distance between the Sun and Pluto is about 39.5 times that separating the Sun and the Earth.

The semi-major axes of the satellites range from 9,378 km to 48.4 million km. The semi-major axes of the innermost, largest, and outermost satellites are not the same across all planetary systems. For instance, the semi-major axis of the Moon (384,400 km) is higher than that of the innermost satellite of any other planet in the Solar

Nathanael-Israel Israel: Historic Discoverer of the Formula to Accurately Decode the Origin of the Universe, of Life, and of Chemicals in a Few Days

System. Later in this book, I will clearly explain the crucial role that the semi-major axis plays in decoding the origin of the universe.

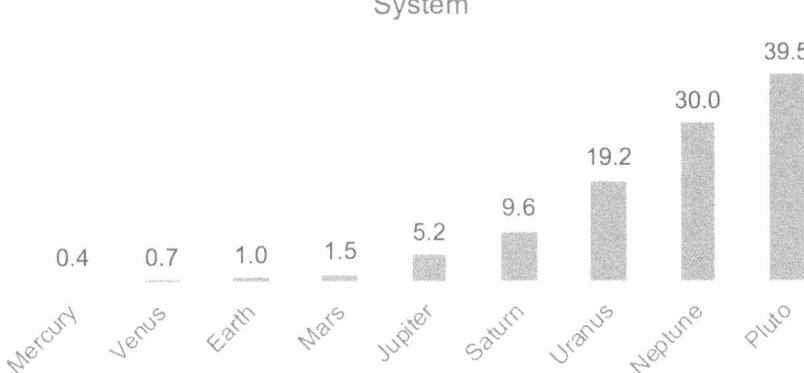

Fig. 2: Semi Major Axis (AU) of Planets in the Solar System

2.3. Orbital speed

Here, you will begin to discover why the orbital speeds of celestial bodies are crucial to properly decrypting the universe's origin. Indeed, every celestial body in the universe is moving. For instance, although not everyone may know that the Earth is moving, everyone can testify that the Sun, the Moon, the stars, and the planets in the sky move every day and night. It took centuries of research before human beings could understand that the Sun's movement across the sky is not because the Sun is moving around the Earth, but because the Earth is moving around the Sun. When the Earth faces the Sun, it is day; when the Sun is on the other side of the Earth, it is night. The universe is filled with other things that move but which we cannot see with the naked eye because they are too small or too distant. The orbital motion or revolution of celestial bodies is their movement around their primary body. For example, in the Solar System, a planet's movement around the Sun is an orbital motion. For instance, the Earth orbits the Sun once every year. The orbital speed of a celestial body is the speed at which it completes an entire revolution or orbit around its primary body. For instance, the orbital speed of a planet in the Solar System is the speed at which it orbits the Sun. Likewise, the orbital speed of a satellite is the speed at which it orbits its primary planet.

The orbital speed of celestial bodies varies according to their position in their orbit. The orbital speed I will be dealing with in this book is the mean orbital speed, which is the average orbital speed of the bodies. Although in my book *"Turbulent Origin of the Universe"*, I devoted dozens of pages to the orbital speed of more than 400 celestial bodies, here, for the sake of space and relevancy, I will highlight just one crucial trend:

the relationship between the orbital speed and the semi-major axis (i.e., distance separating the celestial bodies from their primary body).

Indeed, the orbital speeds of the bodies I studied in the Solar System range from 0.123 km/s to 47.36 km/s. The orbital speeds of the planets in the Solar System range from 4.67 km/s to 47.36 km/s, with Pluto (the outermost planet, now classified as a dwarf planet) the slowest and Mercury (the innermost planet) the fastest. The orbital speed of the Earth is 29.78 km/s. The orbital speed of the Sun (19.4 km/s) is smaller than that of the 4 innermost planets in the Solar System (Mercury, Venus, Earth, and Mars), but higher than that of the 4 giant planets (Jupiter, Saturn, Uranus, and Neptune). The orbital speed of the 210 satellites known in the Solar System as of 2022 varies between 0.123 km/s and 31.58 km/s (Fig. 3).

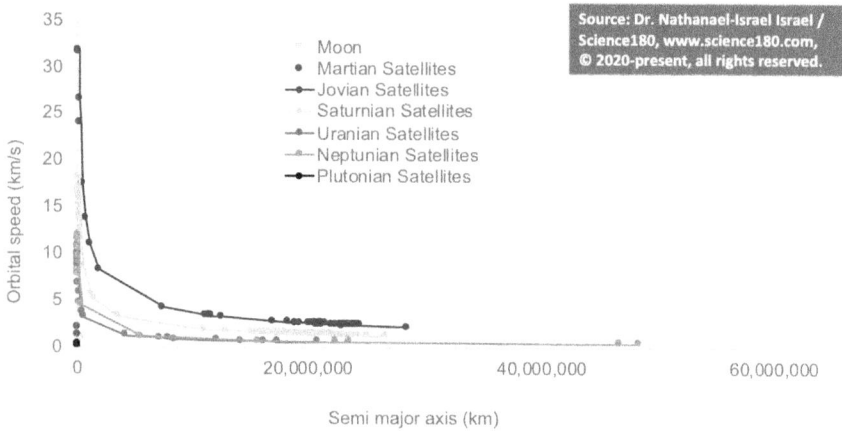

Fig. 3: Orbital speed of the satellites according to their types and semi major axis

Considering the data collected on celestial bodies, I found that the orbital speeds of asteroids and planets in the Solar System decrease as their semi-major axes (i.e., distances from the Sun) increase. In other words, among the bodies orbiting the Sun, the innermost body (meaning the closest to the Sun) is the fastest, while the outermost body (meaning the farthest from the Sun) is the slowest, and the orbital speed decreases from the innermost body to the outermost body. Likewise, the orbital speed of the satellites decreases as their semi-major axes (i.e., average distances from their planet) increase. For all the types of satellites in the Solar System, this trend stands. In other words, when you take the Jovian satellites, their orbital speed decreases from the innermost Jovian satellite to the outermost satellite. Likewise, the orbital speeds of the Saturnian satellites decrease from the innermost to the outermost. The same trend holds for the Uranian, Neptunian, and Plutonian satellites. In other words, in each planetary system, the innermost satellite is the fastest, while the outermost satellite is the slowest.

Nathanael-Israel Israel: Historic Discoverer of the Formula to Accurately Decode the Origin of the Universe, of Life, and of Chemicals in a Few Days

Considering what I said above about the trends of the orbital speed of the planets, asteroids, and satellites, as a rule, in a system of bodies, the orbital speed of the secondary bodies decreases as their semi-major axis (i.e., distance separating the secondary bodies from their primary body) increases. Before I answer later, please think about these questions:

- Why is Mercury (the innermost planet in the Solar System) faster than any other planet in the Solar System?
- Why is Pluto (the outermost planet in the Solar System) slower than any other planet in the Solar System?
- Why is it that in each planetary system, the innermost satellite is the fastest and the outermost satellite is the slowest?
- Why, in general, does the speed of the secondary bodies in a system of bodies decrease from the innermost secondary body to the outermost body?

Very soon, you will see how critical the orbital speed is and why, without it, it may be impossible to ever break the code of the universe.

2.4. Rotation period

In addition to orbiting their primary bodies, celestial bodies also rotate. The rotation period of a celestial body is the time required for it to complete one rotation with respect to a reference, usually a rotational axis or a center. For instance, in 24 hours, the Earth completes one rotation around its rotational axis. I studied the rotation periods of more than 200 bodies in the Solar System, and they range from 3 minutes to 243 days. The smallest rotation period was noted on an asteroid, whereas the highest rotation period was with Venus (the second closest planet to the Sun). More than 75% of the bodies I studied in the Solar System complete their rotation in less than 24 hours. The rotation period of the Sun is about 27 days at the equator.

The rotation period of the planets in the Solar System varies between 9.93 hours and 243 days. Jupiter has the smallest rotation period, whereas Venus has the highest. All the 4 giant planets (Jupiter, Saturn, Neptune, and Uranus) complete a full rotation in less than a day. The rotation period of the satellites ranges from 2.7 hours to 79.32 days. More than 60% of the satellites in the Solar System complete an entire period in less than 24 hours. Here, the main information that I would like for you to remember about the rotation period of the celestial bodies is that, in each planetary system, the rotation period of the satellites increases from the innermost satellite until a satellite I termed the outermost satellite in turbulence Zone 3 (usually located a few satellites outward of the largest satellite in most planetary systems) and then it drops. This means that, in each planetary system, from the innermost satellite to at least the largest satellite in that system, the rotation period increases as the semi-major axis (i.e., distance from the primary planet) increases. Later, I will talk more about turbulence Zone 3, the region where the largest satellites are found in most planetary systems. By the way, the increase in the rotation period of the celestial bodies is associated with a decrease in their rotation (particularly in what is called the rotational angular speed), so these satellites take longer to complete their rotation. In other words, from the innermost satellite to the outermost satellite in turbulence Zone 3, the rotational angular speed decreases, then jumps afterward. As I end this segment, I would like for you to be thinking about this: "Why does the rotational angular speed of the satellites decrease from the innermost satellite until the outermost satellite in Zone 3, and then jump afterwards?"

2.5. Escape velocity

Because of their gravity, celestial bodies in the universe can also "attract" certain things that are close to them or that are in their "gravitational field". For instance, when most objects are thrown into the air, they usually fall back due to the Earth's gravity, but if a certain speed is communicated to them, they can escape the gravitational pull of the Earth. Hence, when spaceships are launched from the Earth, the speed communicated to them is strong enough to prevent the Earth's gravity from pulling them back. NASA and other space agencies have proven that the escape velocity of a celestial body is the *"minimum velocity required at the surface to escape a body's gravitational pull"*. In other words, if the escape velocity is communicated to any object, it can escape the gravity of the celestial body where it was. NASA reported that the escape velocity of the Sun is 617.6 km/s and that of the Earth is 11.19 km/s.

Considering the definition of the escape velocity, if the precursors of the bodies orbiting the Sun (e.g., planetary systems and asteroid systems) were launched with the escape velocity of the Sun from a position near the surface of the precursor of the Sun, they could have escaped the precursor of the Sun. Likewise, if the precursor of the satellites were ejected or launched at the escape velocity of their primary planets from a position near the surface of the precursor of these primary bodies, they could have escaped the precursor of their primary planet. Escape velocity is one of the variables that helped me explain how the precursors of some celestial bodies could have escaped others during the formation of the universe.

2.6. Eccentricity

The orbit of a celestial body is the curved path it follows around its primary body. All orbits are not shaped the same way. While some are circular, others are ellipsoidal. Eccentricity is a parameter that measures how round or elongated an orbit's shape is (Fig. 4).

Fig. 4: Eccentricity of the orbit of celestial bodies

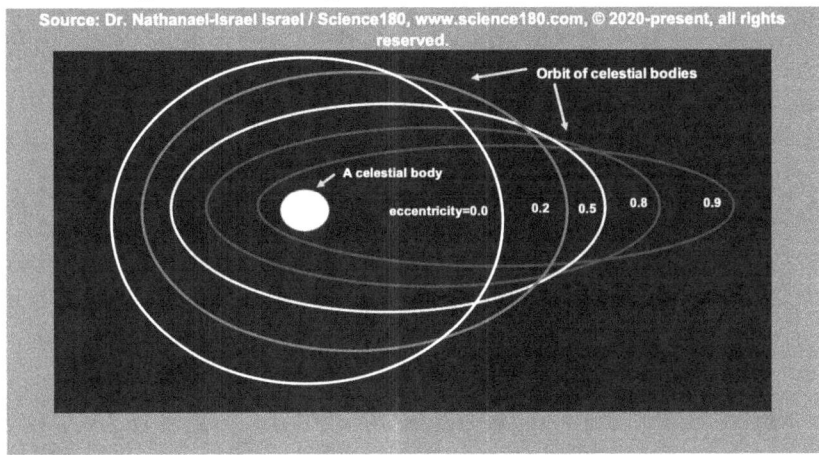

In this segment, I would like to emphasize the trends in satellite eccentricity. Indeed, NASA's scientific data showed that the orbits of most satellites located beyond the largest satellite in most planetary systems are more eccentric than those of satellites located within it. In other words, in most planetary systems, from the innermost satellite to about a few satellites beyond the largest satellite, the orbit of most satellites is almost circular. Beyond the largest satellite in most planetary systems, satellite orbits are usually elongated. Why, before the largest satellite in each planetary system, is the orbit of most satellites nearly circular, but beyond, the orbit is usually very elongated? As I tie things together later, I will address that question.

2.7. Axial tilt

The rotational axis of the celestial bodies (i.e., the axis around which they rotate) is not always straight, but is inclined differently. Axial tilt is a physical parameter that defines the angle between a celestial body's rotational and orbital axes. When the axial tilt of a celestial body is less than 90°, the motion of that body is said to be prograde. In contrast, when the axial tilt is over 90°, the motion is retrograde. The axial tilt of the Sun is 7.25°. The axial tilt of the planets in the Solar System ranges from 0.034° to 177.36°, with the smallest value recorded on Mercury and the highest value on Venus. The three highest axial tilts recorded on planets were observed for both an inner planet (e.g., Venus) and two outer planets (e.g., Uranus and Pluto), suggesting that axial tilt cannot be explained solely by the speed or distance separating these planets from the Sun. Venus (the second innermost planet in the Solar System) is even more tilted than Pluto (the outermost planet) and Uranus (a giant planet). As presented in Fig. 5, the axial tilt of the planets looks as if something had pushed or pulled the planets forward or backward. Looking at this graph reminded me of how a bowler can impart spin and curve to a ball by releasing it from his hand. Since the day I first looked at that graph, I felt like the axial tilt of the planets must relate to the process that formed the bodies.

Fig. 5: Axial tilt of the planets in the Solar System

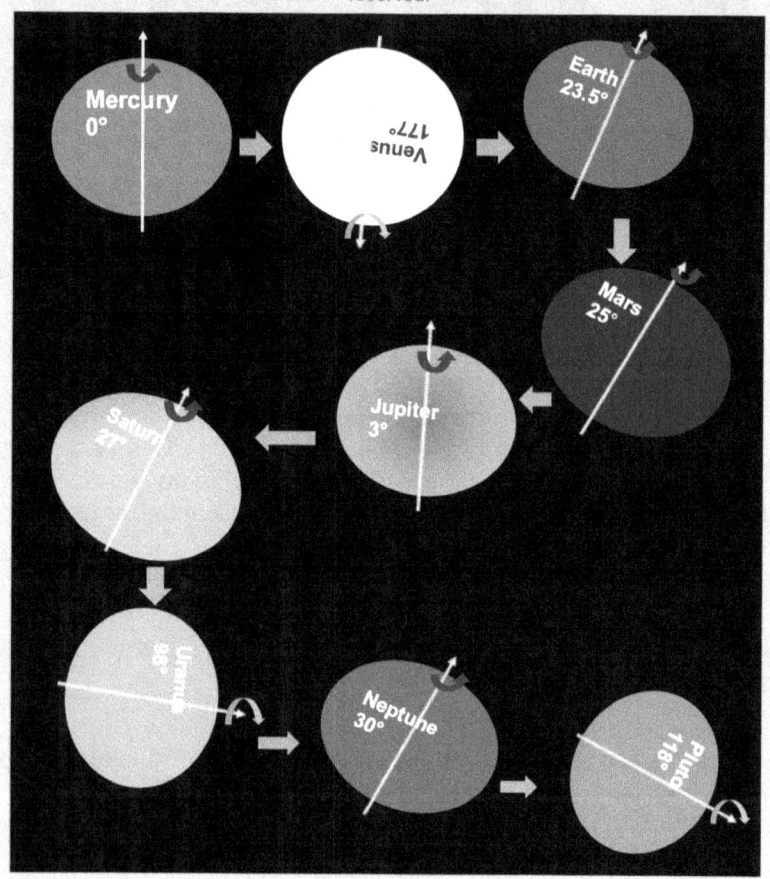

2.8. Orbital inclination and motion

What can we learn from the shocking history of the orbital inclination of the celestial bodies that are "struggling" to properly move in space? The movement of celestial bodies can be defined by a trajectory. But unlike how you can roll a ball on the flat surface of a floor, celestial bodies do not move on a flat surface, or a flat plane, but on a tilted orbital plane. Just as the inclination of hilly roads is different, so also the inclination of the orbit of celestial bodies is distinct and is usually defined by a plane. The orbital inclination is a parameter that describes the tilt of the orbit with respect to

Nathanael-Israel Israel: Historic Discoverer of the Formula to Accurately Decode the Origin of the Universe, of Life, and of Chemicals in a Few Days

a reference. For instance, the plane of the Earth's orbit (also called the elliptic plane or the elliptic) is the reference plane used to define the orbital inclination of the planets and asteroids in the Solar System. This implies that Earth's orbital inclination is 0. Here, I would like to mention that the Earth's orbital plane is tilted 7.25° with respect to the Sun's equator. As for the satellites, the reference plane used to measure the orbital inclination is the equator of their primary planet. In other words, the orbital inclination of a satellite is the angle between its orbital plane and the equator of its primary planet.

Based on NASA's raw data, I studied the orbital inclinations of more than 400 celestial bodies in the Solar System. Instead of delving into their details here, I would like to focus on the positions of the satellites with the highest orbital inclinations. Indeed, as you will see in Fig. 6, the outer satellites usually have the highest orbital inclination, whereas the innermost satellites frequently have the smallest orbital inclination. In general, from the innermost satellite to the largest satellite in each planetary system, the orbital inclination of the satellites is usually negligible. In contrast, beyond the largest satellite in each planetary system, the highest orbital inclinations are usually found.

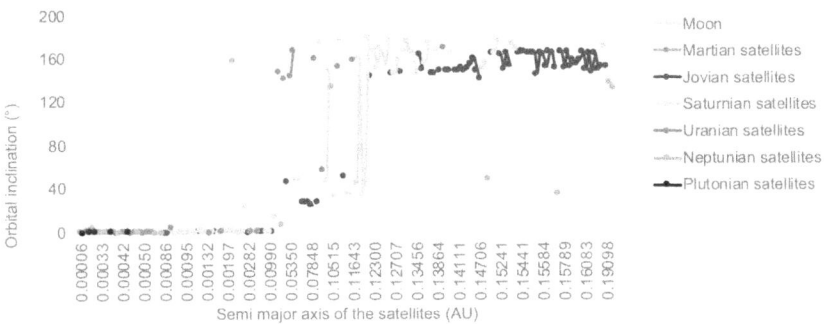

Fig. 6: Distribution of the orbital inclination of the satellites according to their types and semi major axis

Most of the planets in the Solar System orbit the Sun in the same direction as the Sun's rotation, which is counterclockwise when viewed from above the Sun's North Pole. The orbital inclination is also used to define the motion of the celestial bodies as prograde, retrograde, or chaotic. Indeed, when the orbital inclination is between 0° and 90°, the motion is prograde, meaning that the body and its primary body are rotating in the same direction. When the orbital inclination of a body is between 90° and 180°, the motion is retrograde, giving the impression that the body is rotating in a reverse direction. For instance, a retrograde planet orbits its primary planet in a direction opposite to that of the planet's rotation. Sometimes, a few prograde satellites can be found between retrograde satellites, and this usually happens beyond the largest satellite in some planetary systems. Finally, a chaotic rotation is the irregular and unpredictable rotation of a celestial body that does not usually have a fixed rotational axis or period. Some satellites with chaotic motions are so close to one another that they sometimes share the same orbit. Below, I sketched the orbit of celestial bodies (Fig. 7). I would like to end this chapter with two questions that caught my attention, and that I will answer later:

- Why, in each planetary system, is the orbital plane of the satellites, ranging from the innermost satellite to about the largest satellite, not usually inclined?
- Why is the orbital plane of most of the satellites located beyond the largest satellite in most planetary systems usually very inclined, as if something flipped it over?

Fig. 7: Prograde and retrograde orbit of celestial bodies

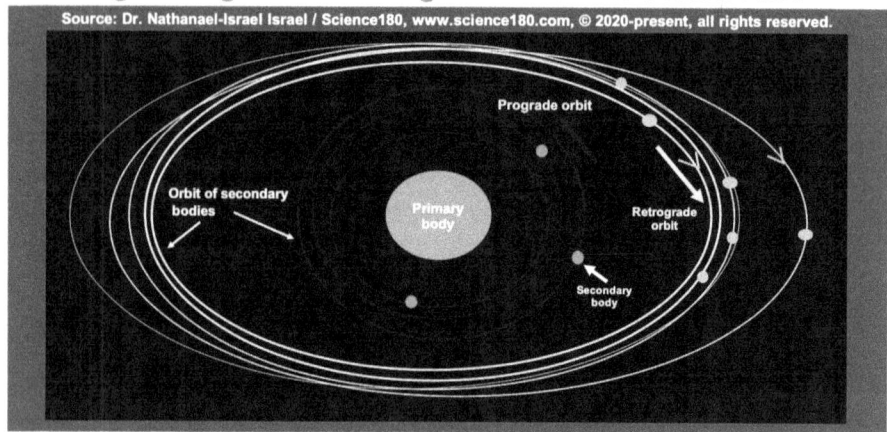

2.9. Radius

Radius is the main variable used to describe the size of the celestial bodies. The radius of a spherical object is the distance from its center to its outer edge. In this book, wherever I talk about radius, I mean the equatorial radius, which is the radius at the equator of the bodies.

I investigated the radius of more than 400 celestial bodies in the Solar System, and of them all, the Sun is the largest, having a radius of 696,000 km. The second largest body in the Solar System is Jupiter, and its radius (71,492 km) is 10.3% that of the Sun. The radius of the Sun is about 109 times that of the Earth (6378.14 km). The radii of the satellites I studied in the Solar System range from 300 meters to 2634.1 km. The radius of the Moon is 1738.1 km. As illustrated in Fig. 8, most of the largest satellites are not the innermost or the outermost satellite, but are located some distance in between. The semi-major axis of the largest satellite is not the same for all planetary systems. Finally, although small satellites are found among larger ones, the outermost satellites are usually the smallest. The main point I want to emphasize about the radius is the location of the largest satellites and that, beyond them, most of the satellites are very small.

Nathanael-Israel Israel: Historic Discoverer of the Formula to Accurately Decode the Origin of the Universe, of Life, and of Chemicals in a Few Days

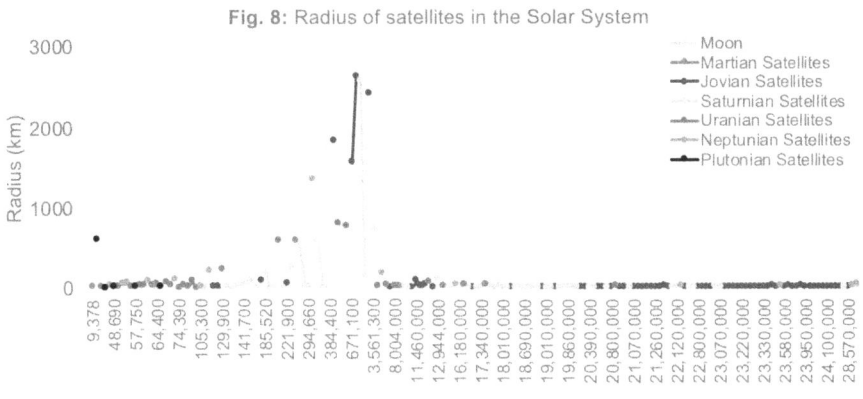

Fig. 8: Radius of satellites in the Solar System

2.10. Density

Density is the ratio between mass (or weight) and volume (amount of space occupied by an object). Because mass is usually expressed in kilograms (kg) and volume in cubic meters (m^3) or multiples thereof, the international unit of density is kg/m^3. Density measures the degree of compactness or consistency of a body. I studied the density of more than 150 celestial bodies in the Solar System. It ranges from 250 kg/m^3 to 6730 kg/m^3. Many of the lowest densities were observed among the Saturnian satellites, with Saturn itself among the least dense bodies in the Solar System. The densest bodies were found among the main-belt asteroids and planets. Although the Sun is the largest body in the Solar System, its density (1408 kg/m^3) is among the smallest. Indeed, about 58% of the densities I studied exceed that of the Sun. This suggested that the Sun's precursor was less compressed than that of other bodies in the Solar System.

The density of the planets in the Solar System varies between 687 kg/m^3 and 5514 kg/m^3, the highest value being recorded with the Earth, whereas the smallest value was with Saturn (Fig. 9). The terrestrial planets (i.e., Mercury, Venus, Earth, Mars, and Pluto) are denser than the Sun. The least dense planets are the 4 giant planets (i.e., Jupiter, Saturn, Uranus, and Neptune). Among the satellites, the Moon's density (3344 kg/m^3) is the second-highest. The densest satellite is a Jovian one. The densest satellite in each planetary system is found in the most turbulent zone, suggesting that turbulent processes may have contributed to defining the density of the celestial bodies. The key information I would like to underline in this section on density is that the largest celestial bodies are generally less dense than the smaller ones. Hence, the Sun and the giant planets are not the densest bodies in the Solar System.

Fig. 9: Density (kg/m³) of the Sun and the planets in the Solar System

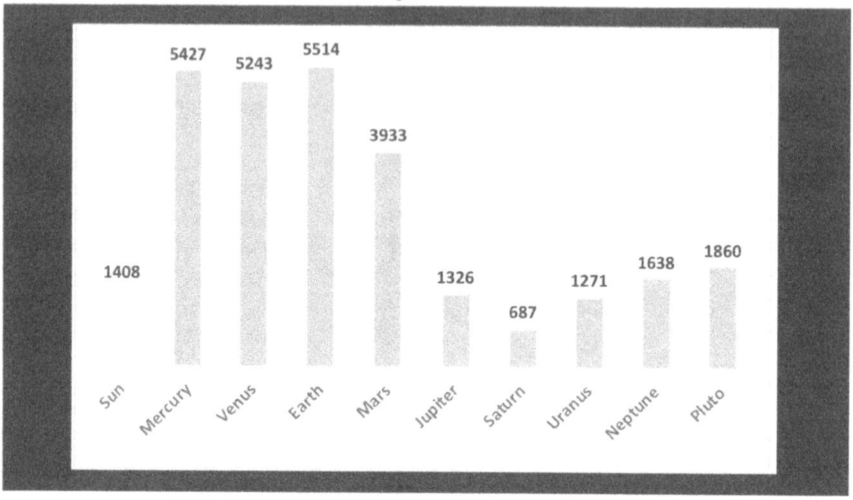

2.11. The 99% versus 1% rule of the system-additive variables

One of the questions that preoccupied me during my investigation of the formation of the universe was the law that can explain the fragmentation of the precursors of the celestial bodies into their respective daughter bodies. To address this issue, I initially wrote about 100 pages on what I termed the system-additive variables, but here, without going into detail, I will present a quick summary of 3 of them (e.g., mass, kinetic energy, volume) that I think may interest the public. I calculated the kinetic energy of the celestial bodies (the energy they possess due to their motion) using their speeds and masses. Using raw data on the celestial bodies in the Solar System, I showed that more than 99% of the mass, kinetic energy, and volume of the bodies in the Solar System is in the Sun, while less than 1% is in the bodies orbiting the Sun. Likewise, I proved that more than 99% of the mass, kinetic energy, and volume of the bodies in the planetary systems is found in the primary planets, while less than 1% is found in the satellites. As a rule, the scientific evidence indicates that more than 99% of the mass, kinetic energy, and volume of the bodies in a system are in the primary body, whereas less than 1% is in the secondary bodies (Israel 2025a).

Moreover, when I considered the bodies orbiting the Sun, I found that more than 99% of their mass, kinetic energy, and volume are in the giant planets, while less than 1% is in the other bodies. Likewise, when I considered the satellites, I realized that more than 99% of their mass, kinetic energy, and volume are in the satellites located in turbulence Zone 3 (where the largest satellites are usually found). Therefore, as a rule, more than 99% of the mass, kinetic energy, and volume of the secondary bodies in a system of secondary bodies is found in the turbulence Zone 3, whereas less than 1% is found in the other secondary bodies (Israel 2025a). In closing, here are 3 questions that I will be answering after going over a few more facts:

Nathanael-Israel Israel: Historic Discoverer of the Formula to Accurately Decode the Origin of the Universe, of Life, and of Chemicals in a Few Days

- Why is more than 99% of the mass, kinetic energy, and volume of the bodies in the Solar System found in the Sun and less than 1% found in the bodies orbiting the Sun?
- Why is more than 99% of the mass, kinetic energy, and volume of the bodies in the planetary systems found in the primary planets and less than 1% found in the satellites?
- Why is more than 99% of the mass, kinetic energy, and volume of the satellites in the planetary systems found in the most turbulent zone (usually Zone 3), and less than 1% found in the other zones?

2.12. Breakup, stratification, and squashing of fluids

A liquid (like water) and a gas (like air) are examples of fluids. Fluid breakup is a phenomenon that fragments a mass of fluid into many smaller fluid masses. It has been shown that in nature, fluids break into sequential cascades (Eggers and Villermaux, 2008), meaning a larger fluid mass breaks into smaller ones, which in turn can break into much smaller ones, and so on until the fluid can no longer break. It was also shown that a fluid's viscosity and turbulence can affect its breakup.

When fluids are stacked, they can form layers. There are many types of fluid flow. In a shear flow, fluid layers are laterally displaced relative to each other. Rheology is the branch of physics that studies the interrelations between forces and deformations of matter in a fluid flow. Rheological studies have shown that fluid layers in a shear flow slide over one another, with each layer moving faster than the one beneath it (Malvern, 2016). In other words, when a stack of fluid layers is moving, the top layers move faster than the bottom layers. Another study showed that when a fluid is squashed, its rotation slows, and when it is stretched, its rotation increases while its diameter decreases (Price, 2006; George, 2013; Petitjeans and Bottausci, 2020).

Other studies have shown that when fast-moving fluid layers interact with one another, instability can emerge at their interface and develop into turbulence, a state of matter characterized by the formation of vortices, which are 3-D structures that can rotate. Therefore, when fluid layers in a stack move quickly past one another, turbulence can occur, affecting the characteristics of the structures that form from them.

As I wrap up the 12 facts I planned to present in this chapter, I would like to say that one of the foundations of my demonstration of the universe's origin is my deep insight into the turbulence that occurred at the beginning. To facilitate my study of the satellites according to their general characteristics, I divided them into regions or groups that I termed turbulence zones based on the summary of their main characteristics:

- Turbulence Zones 1, 2, and 3, which satellites have the smallest orbital inclination and the smallest eccentricities. Turbulence Zone 1 is where the innermost and fastest satellites are usually found. The rotational period of the satellites increases from the innermost to the outermost satellite in Zone 3, then increases again. In other terms, a key characteristic of the outermost satellite in Zone 3 of each planetary system is that it has the highest rotation period, meaning that they took more time to finish one rotation around its rotational axis. Turbulence Zone 3 is where the biggest satellites of most planetary systems are usually found.

- Turbulence Zones 4 and 5 are located outward of Zone 3 and are usually dominated by satellites having the highest orbital inclinations, the highest eccentricities, and the smallest radius. Zone 5 is where the outermost and slowest satellites are found in the 4 giant planetary systems.

Given the importance of these turbulence zones and the journey that led me to their discovery, I devoted a separate book to them. The turbulence zones allowed me to study the bodies based on the intensity of the turbulence their precursors experienced.

When fluids are flowing, they are believed to be pushed and pulled by surrounding fluid parcels (a fluid parcel is a group of fluid or a fluid patch) according to the stresses or forces created by the flow. In other words, the motion of a fluid parcel is better predicted not in isolation from its surroundings, but by also considering the motion of the surrounding fluid parcels. When applied to the precursors of fluids during the formation of the universe, to properly study the motion of a fluid layer, attention must be given to the fluid layers adjacent to it, namely the upstream or top layer and the downstream or bottom layer in its immediate surroundings. Because all we have today are celestial bodies, not fluid layers, to study the process involved in the motion of their fluids of their precursors, attention must be given to the bodies upstream and downstream of each body of interest.

2.13. Postulates

So I can easily refer to them as in the remaining chapters, I summarized the key facts I summarized above into postulates, which are going to drive my theory:

- Postulate 1: At the beginning of the universe, a bulk of an origin particle mysteriously appeared, and then, under the influence of a turbulent process, it was fashioned into the celestial bodies present in the universe today.
- Postulate 2: The precursors of the celestial bodies in the universe were fragmentated and gathered together not by chance, but according to a sequential cascade of breakups following precise rules.
- Postulate 3: The turbulence intensity and the viscosity of the fluids of the precursors of celestial bodies affected their breakup and the nature of their daughter bodies.
- Postulate 4: The universe is a nest of primary bodies usually orbited or surrounded by secondary bodies.
- Postulate 5: The precursor of the secondary bodies of a system of bodies escaped the precursor of their primary body at about the escape velocity of that primary body.

- Postulate 6: During the fragmentation of the precursor of a system of celestial bodies, more than 99% of the mass, kinetic energy, and volume went into the precursor of the primary body, while less than 1% went into the precursor of the secondary bodies.
- Postulate 7: The orbital speed and the semi-major axis of the secondary bodies in a system of bodies are negatively correlated.
- Postulate 8: The rotational period of the satellites increases from the innermost satellite to the outermost satellite in turbulence Zone 3, then it suddenly decreases.
- Postulate 9: In the planetary systems, most of the satellites with the highest eccentricities, the highest orbital inclinations, and the smallest radius are usually located outward of the largest satellites.
- Postulate 10: In a moving set of fluid layers, the speed of the layers decreases from the top to the bottom, and the squashing of the top layers stresses the bottom layers, causing them to stretch, spin up, and form smaller bodies.

2.14. Take-home message

To recapitulate, the facts I enunciated in this chapter show that the universe is organized as sets of primary bodies surrounded by, or orbited by, secondary bodies for a reason. The orbital speed of the secondary bodies decreases as their semi-major axis (i.e., distance from their primary body) increases. Although celestial bodies tend to attract things in their gravitational field, when the escape velocity is communicated to a body, the latter can escape the gravitational pull of the celestial body that has that escape velocity. The orbits of celestial bodies are inclined and elongated in different ways. Most of the satellites with the highest eccentricities, the highest orbital inclinations, and the smallest radius are located beyond or outward of the largest satellite in most planetary systems. In contrast, the largest satellites have the smallest eccentricities and orbital inclinations. Furthermore, while the rotational axis of the celestial bodies is tilted as if these bodies were pushed or pulled, that of others is not. More than 99% of the mass, kinetic energy, and volume of the celestial bodies of the system of celestial bodies in the Solar System is found in the primary bodies, while less than 1% is in the secondary bodies. In general, just like the Sun, the giant planets in the Solar System are less dense than the smaller planets (i.e., terrestrial planets). Finally, in a moving stack of fluid layers, the layers' speed decreases from the top to the bottom, and squashing or compression at the top stresses the bottom layers, causing them to stretch, spin up, and form smaller bodies. I summarized the 12 facts as postulates that I will refer to frequently in the rest of this book. At this point, I will begin explaining the underlying causes of the trends I presented in this chapter.

Another Book by Nathanael-Israel Israel:
TURBULENT ORIGIN OF THE UNIVERSE

THE FIRST AND ONLY SCIENTIFIC BOOK THAT ACCURATELY EXPLAINS EVERYTHING YOU NEED TO UNCONVENTIONALLY, EASILY, AFFORDABLY, AND ENJOYABLY DECODE THE UNIVERSE FORMATION

In *"Turbulent Origin of the Universe"*, filled with great diagrams and digestible scientific facts, you will discover, learn, or get:

- The all-in-one, proven & uncomplicated scientific formula that accurately decoded the formation of the universe, and that explained the birthdate of the stars, planets, satellites, asteroids, and all other celestial bodies in the universe, so you can position yourself to stay on top of your competitors, avoid repeating crucial mistakes that many people have ignorantly made at their own perils
- Extraordinary, unprecedented, accurate insights into the first factors (e.g. early universe physics) that defined the history and formation of the universe so you can tap into deep scientific secrets you ignore, and set yourself apart from others
- The 4 simple things without which it is impossible for anyone to ever understand the formation of the universe, think accurately, work differently, achieve, or perform better for superior results
- The verified key to move the cosmological mountains of misunderstanding, so you can confidently free your mind from doubts, improve your health, and prevent you from any danger connected with sticking with wrong assumptions
- Save time and money, and enjoy your life once you remove errors holding your true understanding of the universe-origin captive
- Historic scientific proof of whether a planet was formed in 2.82 days, whether a satellite was formed in 3.32 days, and whether a star was formed in 3.69 days after the beginning of the universe; so you can creatively produce and address a broader work spectrum by learning how to effectively communicate with and establish unusual connections between otherwise disconnected and disparate scientific data
- The scientific formula that successfully tested the existence of God in a way that shocked believers, skeptics, and all other freethinkers

Nathanael-Israel Israel: Historic Discoverer of the Formula to Accurately Decode the Origin of the Universe, of Life, and of Chemicals in a Few Days

- Why the scientific community has failed to sufficiently explain the origin of the universe; and understand how existing theories have missed and undefined central ideas, and imposed limits on the vision of scientists
- Specific in-depth knowledge, up-to-the-minute information, and ideas so you can expand your market, cut useless costs, stop wasting time on inadequate projects, and start focusing on the profitable solutions (Science180.com/scientific)
- How Science180 Academy can strategically enlighten you, guide you to navigate and filter the massive data collected on the universe, so you can answer the world's most challenging questions, remove any scientific and philosophical cataracts that may be blocking you, and bring you many steps closer to your best life
- How to better resonate with your target market that is craving something original that breaks wrong explanations of the universe-origin

Get *"Turbulent Origin of the Universe"* today to begin an incredible journey of accurately decoding the universe and change your life forever!

Dr. Nathanael-Israel Israel is told by people that he is the #1 Universe-origin, Life-origin, and Chemicals-origin Expert. He is the founder of Science180 and the author of many books on the origin of the universe and its content. To learn more about how he may help you, visit Israel120.com.

Nathanael-Israel Israel: Historic Discoverer of the Formula to Accurately
Decode the Origin of the Universe, of Life, and of Chemicals in a Few Days

CHAPTER 3

WILL THIS ORIGINAL ANALYSIS RADICALLY REVIVE THE ABANDONED DREAM OF THE FATHERS OF CLASSICAL MECHANICS WHILE BURYING THE HOPE OF THE RELATIVISTS?

Considering the facts I presented in the previous chapter, you may be eager to know the underlying factor behind them and the story that properly supports them. The goal of this chapter is to give you a quick glimpse at some aspects of the demonstrations I will be doing in the rest of this book. Beforehand, I will go over some postulates I presented in the previous chapter and show how they led me to unravel the process by which the universe formed.

3.1. What explains the trends in the orbital speed of celestial bodies

I will start this demonstration by first referring to postulate 10, which says that "In a moving set of fluid layers, the speed of the fluid layers decreases from the top layer to the bottom layer and the squashing of the fluids stresses the bottom layers and can cause them to stretch, spin up, and form smaller bodies". As I reviewed this postulate carefully, I understood that if the speed of the fluid layers in a stack can decrease from the top layer to the bottom layer, it also implies that if these fluid layers can be gathered together to form celestial bodies, the speed of the bodies will also decrease from the top layers to the bottom layers. In other words, celestial bodies formed from the top fluid layers would move faster than celestial bodies formed from the bottom layers. In my book *"Turbulent Origin of the Universe"*, I showed that the top layers formed the innermost bodies, while the bottom layers formed the outermost bodies. This caused the speed of the bodies formed from the stack of fluid layers to decrease from the innermost to the outermost. Furthermore, because postulate 10 also says that the bottom layers will be stretched, spin up faster, and thin (leading to the formation of small bodies), I also deduced that the bodies formed from the bottom layers would rotate faster than the bodies formed from the layers that squashed them, and they would be small. Furthermore, because the stretching of the layers could have stretched their orbit, I deduced that the orbits of the bodies formed from the bottom fluid layers would be elongated and inclined. In other words, postulate 10 implies that, when bodies are formed from a stack of moving fluid layers, the top layers squash the bottom layers and, in the end, the bodies formed from the top layers move faster than the bodies formed from the bottom layers; and the bodies formed from the bottom layers are small, and their orbits are elongated. The implication of postulate 10 is exactly the characteristics that the celestial bodies displayed in postulates 7, 8, and 9 (see the previous chapter):

- Postulate 7: The orbital speed of the celestial bodies decreases from the innermost body to the outermost body.
- Postulate 8: The rotation angular speed of the satellites decreases from the innermost satellite to the outermost satellite in turbulence Zone 3, then it suddenly jumps, meaning that beyond the largest satellites, the rotation of the satellites spins up; and
- Postulate 9: Most of the satellites with the highest eccentricities, the highest orbital inclinations, and the smallest radius in the planetary systems are usually found outward of the largest satellites.

Considering the demonstration I did above, I deduced that, at one point in their formation, the satellites' precursors were organized as fluid layers that moved at different speeds. Fluids are not a solid that always moves as a block; hence, some that are on top can move past those that are beneath them. Because the top layers squash the bottom layers, the fluid's speed decreases from top to bottom. During the process, the upper fluid layers were collected into the innermost satellites, while the lower layers were used to form the outermost bodies. Because the fluid-layer speeds imparted to the bodies formed from them carry over to the bodies, the trends in the fluid-layer speeds are reflected in the satellites' speeds. Hence, the orbital speed of the satellites decreases from the innermost to the outermost, with the innermost having the highest and the outermost the lowest. Furthermore, the decrease in the satellites' orbital speed also reduced their rotational angular speed. But, as postulate 10 explained, the increase in rotation and the small size of the satellites beyond the largest ones can be explained by how the squashing of the fluid layers of the precursor to the largest satellites could have affected bodies located beyond it. Indeed, postulate 10 says that when a fluid layer is squashed, its rotation increases, and the radius of the bodies formed from it can also be small. In other words, the flattening and stretching of the structures formed in the bottom fluid layers reduced their size. Hence, most satellites located beyond the largest ones have smaller radii. Likewise, the squashing, squeezing, or compressing of the fluid layers also stretched them, causing the orbits of the bodies formed from them to elongate. Hence, the orbit of the satellites located beyond the largest ones is usually elongated. In the process, the precursors of some satellites were flipped over, therefore increasing their orbital inclination as mentioned

Nathanael-Israel Israel: Has had the Honor to be Acknowledged the First Human Being that Scientifically Reconciled Science and Biblical Creation

in postulate 9. Satellites located inward of the largest ones usually have the smallest orbital inclination and eccentricity because their precursors were not squashed by the precursors of the largest satellites. To summarize, postulates 7, 8, 9, and 10 imply that, during the formation of the satellites, fluid layers formed, and their interactions explain the trends observed in the satellites' orbital speed, rotation, orbital inclination, and eccentricity (orbit elongation). I will revisit eccentricity and orbital speed in a later segment.

Using the above logic, I explained that the decrease in the orbital speeds of the planets and asteroids as their semi-major axes (i.e., distances from the Sun) increased was caused by how the fluid layers of their precursors were organized and how they affected one another. Indeed, during the formation of the planets and asteroids in the Solar System, the upper fluid layers squeezed the lower layers. The top fluid layers formed the innermost planets and asteroids, while the bottom layers formed the outermost . By the time they were formed, the innermost planets and asteroids orbit the Sun faster than the outermost planets and asteroids, and orbital speed decreases from the innermost to the outermost.

To recapitulate how the celestial bodies acquire their orbital speed, it is important to remember that the fluid layers of the precursors of the bodies were stacked one layer on top of the other and slid over one another, with each layer moving faster than the one beneath it. In other words, the fluid layers at the top (the uppermost layers) had the fastest speed, while the bottom layers had the slowest. As the fluid layers split from one another, the top fluid layers formed the innermost bodies (i.e., closest bodies to the primary bodies) while the bottom layers birthed the outermost bodies (i.e., farthest or most remote bodies from the primary bodies). Due to their position in the stack of fluid layers of their precursors, the innermost bodies ended up having the highest orbital speed, while the outermost bodies have the lowest orbital speed. That is why, in a planetary system, the innermost satellites are the fastest, while the outermost are the slowest; likewise, in the Solar System, the innermost planets and asteroids are faster than the outermost.

Now that I had shown that the precursors of satellites, planets, and asteroids once existed as fluid layers, I needed to know the direction of their flow. Knowing how these fluid layers flowed can provide clues about their causes and origins. By tracking the sources of fluid flow across various systems of celestial bodies, I can pinpoint how these bodies were formed. When I considered the trends of the orbital speed (postulate 7), the organization of the universe as a nest of primary bodies orbited by secondary bodies (postulate 4), and how fluid layers of the precursors existed and flowed during the formation of the secondary bodies, I realized (and I extensively proved it in my original book *"Turbulent Origin of the Universe"*) that, in each system of bodies, the fluid layers of the secondary bodies flowed from about the position of the primary body to the position of the secondary bodies. In other words, the precursor of the satellites flowed from about the position of the precursor of their primary planet to the position of the outermost satellites. Likewise, the precursors of the planets and asteroids in the Solar System originated from roughly the position of the precursor of the Sun to that of the Solar System's outermost celestial bodies. Following this line of thought, I generalized that in a stellar system, the fluids of the planets and asteroids could have flown from the position of the primary star, while in a planetary system, the fluids of the satellites could have flown from the positions of their primary planets.

Because the precursor of the secondary bodies flowed from about the position of the precursor of the primary body, it cannot be excluded that, at one point, these precursors were once together as one fluid body. For instance, the precursor of the satellites and of their primary planet was once one body: the precursor of their planetary system. Likewise, the precursor of the Sun and the precursor of the bodies orbiting the Sun were once one body: the precursor of the Solar System. In other words, the precursor of the Solar System was split to birth the precursor of the Sun and the precursor of the bodies orbiting the Sun. Then, as the precursor of the bodies orbiting the Sun moved, it split to form the precursor of the planetary systems, which then split to form the precursor of the planets and satellites. Between the precursors of the planetary system, the precursor of asteroids and their systems were also formed.

Finally, in the Solar System, scientific evidence shows that, when viewed from the North Pole, most celestial bodies orbit the Sun in a counterclockwise direction. This suggests that the fluid layers of the precursors of most bodies orbiting the Sun flowed counterclockwise when viewed from the north pole. In general, the precursor fluid layers flowed in the direction of the force, which pushed them away from their mother's initial position. The organization of the fluids in the precursors of celestial bodies into layers can also explain the presence of some strata in the Earth's crust. For instance, when a well is dug on Earth, soil strata are usually found in layers regardless of location. These layers of matter were formed from the fluid layers of the precursor of the crust. The upcoming graph (Fig. 10) is a sketch of the fluid layers in the precursor of a system of bodies.

Nathanael-Israel Israel: Has had the Honor to be Acknowledged the First Human Being that Scientifically Reconciled Science and Biblical Creation

Fig. 10: Layers of fluids in the precursor of a body or precursor of a system of bodies

3.2. Lesson from the escape velocity of celestial bodies

So far, I have shown that the fluid layers of the precursors of secondary bodies flowed from about the position of the precursor of their primary bodies toward the position of the outermost secondary bodies. As of today, the distance separating the secondary bodies from their primary bodies in the Solar System is well known, and that is what I described in the previous chapter as the semi-major axis. But to determine how long it could have taken for the precursor of the secondary bodies to move from their primary body's position to their current position, I needed to know the speed of that movement. To answer that question, I turned to postulate 5, which states that "the precursor of the secondary bodies of a system of bodies escaped the precursor of their primary body at about the escape velocity of that primary body". Indeed, I already showed that the precursor of the primary body and its secondary bodies were once together before being split. I also showed that the fluids of the precursors of the secondary bodies flowed from about the position of their primary body to the position of the outermost secondary body. For that flow to occur, the precursor of the secondary bodies must have moved away with a speed at least equal to the escape velocity of the precursor of the primary body.

In the previous chapter, I showed that the primary bodies are larger than their secondary bodies. In fact, postulate 6 showed that 99% of the mass, kinetic energy, and volume of the bodies in a system are found in the primary body. In other words, because the primary bodies are much larger than the secondary bodies, for the precursor of the secondary bodies to escape the precursor of the primary bodies, the former must have been moved away at a speed at least equal to the escape velocity of the precursor of the primary body. Otherwise, the precursor of the primary body could have prevented the precursor of the secondary body from escaping. For with a speed lower than the escape velocity, the fluids of the precursor of the secondary bodies could not have escaped the precursor of the primary body. Therefore, the precursor of the bodies orbiting the Sun (e.g. planetary systems and asteroid systems) could have escaped the precursor of the Sun at about the escape velocity of the Sun, while the precursor of the satellites escaped the precursor of their primary planet at about the escape velocity of that planet. As a rule, the precursor of the bodies orbiting a star escaped the precursor of that star at about the escape velocity of that star. Similarly, with a speed near the escape velocity of their primary planet, the precursor of the satellites and rings orbiting a planet could have escaped the precursor of that planet. Later, I will specify the escape velocity of the celestial bodies as I calculate their formation timelines.

3.3. Changes in the fluids of the precursors of bodies

The bodies in the universe were not formed overnight; their precursors passed through stages of development during which they were shaped by a complex, turbulent process. I already explained how the fluid layers in the precursors of celestial bodies interact with one another and affect the characteristics of the bodies born from them, but in this section, I will explain how some events that occurred inside those fluid layers contributed to the formation of the celestial bodies. Indeed, as the precursors of bodies in the universe were being reorganized by turbulence, their surface and interior identities changed. New structures were formed in the fluid layers until a time when major reorganizations or rearrangements were no longer "possible". For instance, as the fluid layers flowed, vortical structures formed. In the process, small vortices were incorporated into larger ones and so on and so forth until all the fluid layers were collected into their corresponding celestial bodies. The vortices that formed inside the fluids played an important role in how the layers were collected. The reorganization of the matters in the fluids of the precursors also depended on the duration of their instability. For instance, a certain volume of coffee mixed with a certain volume of cream can yield different structural mixtures depending on the duration of time and the intensity with which the mixture is stirred or shaken.

Nathanael-Israel Israel: Has had the Honor to be Acknowledged the First Human Being that Scientifically Reconciled Science and Biblical Creation

Likewise, the precursors of the bodies in the universe could have birthed different kinds of daughter bodies if the turbulence they went through was different. To put it another way, the turbulence of the early universe was responsible for the characteristics of the bodies in the universe today. I have extensively demonstrated in my book *"Turbulent Origin of the Universe"* that if the turbulence during the universe's formation had been "calibrated" differently, a different world could have emerged. During the movement of the precursors of the bodies throughout space, phases of "differentiation", including expansion, compression, or squeezing, occurred. The topological changes the precursors underwent altered their identities, properties, shapes, and many other features. In the process, some bodies "inherited" certain properties from their mothers, just as children inherit their DNA from their parents. The features of the precursors could have determined the characteristics (e.g., size and distribution) of their descendants. To illustrate, if the precursor to the Solar System had been launched at a different speed and from a different angle, the Sun and the bodies orbiting it could have been different. Similarly, if NASA plans to send one of its space shuttles to the International Space Station, it needs to calculate the coordinates correctly so it reaches its target. If space shuttles are launched at different speeds, angles, or directions, they can miss their destination.

'Science180 Academy' Success Strategy:
SCIENCE180 ACADEMY OVERVIEW

Science180 Academy is a training, speaking, consulting, and mentoring program designed to groom and empower people of all backgrounds in the truth about the origin of the universe, life, and chemicals. According to their background and interest, trainees are taught different levels of scientific facts to grasp a deeper understanding of the origin of the universe, how to properly think to unearth mysteries hidden in the massive scientific data collected across the globe, but which is unfortunately less analyzed. If you want to be enlightened and equipped so you can cause positive changes in your respective field of expertise, then Science180 Academy program is for you.

Science180 Academy does not confer college credit, grant degrees, or grade its attendants, participants, or students. It is not an accredited university or college, but is the one-stop-destination for universe-origin, life-origin, and chemicals-origin experts. It is where scientists and laypeople get all their origin-related questions properly answered. It is the only place where the accurate interpretation of the universe-origin, life-origin, and chemicals-origin data matters a lot.

Science180 Academy brings together Dr. Nathanael-Israel Israel (the Founder of Science180) and other experts to deliver outstanding value, insight, and lessons to assist you to accurately understand the true origin of the universe, chemicals, and life, so you can tap into that knowledge to improve lives perpetually. Nathanael-Israel's goal is to give you practicable and undeniable proofs of the formation of the universe so you can be fired up to become the best version of you, and to cause positive changes to your initiatives that will profit you today and forever. For Nathanael-Israel, decoding the origin of the universe and everything in it is not a job, but his life mission, and helping others to fully understand that is his mission. Visit Science180Academy.com today to start.

If you are still wondering if Science180 Academy is for you, let me also inform you that some of Science180's clients and prospects have a profound technical knowledge and background in science, while others don't. Some are creationists (e.g. Science180 creationism, Young Earth creationists, Old Earth creationists, Intelligent design proponents), others are anti-creationists. Some are believers, others are freethinkers (including atheists, humanists, rationalists, agnostics, nontheists, nonreligious, skeptics, nonbelievers, religiously unaffiliated, spiritual-not-religious, ex-believers, and doubters). Regardless of their background, belief, or disbelief, Science180 works with each of these people to figure out their needs, priorities, and the products and services that best fit them. Science180 improves their knowledge, experience, performance, and answer their questions (related to the universe-origin, life-origin, and chemicals-origin) by crafting a personalized program that perfectly matches their interests, needs, and things that are dear and meaningful to them whether it is to:

- Become the leader that captures the heart of your followers, prospects, and customers craving for an unconventional explanation of the origin of the universe, life, and chemicals
- Benefit from continual updates and assistance during your journey to decode the universe, and clear your way for the universe-origin related freedom, power, technology, innovation, and breakthroughs of the future.
- Bypass technical knowledge that restricts non-experts from accessing the origin-related truth contained in the massive scientific data, and get to the bottom of scientifically-locked origin-related secrets regardless of your background.
- Challenge the cosmological status quo and embrace the real change that will disrupt the cages that were holding you

Nathanael-Israel Israel: Has had the Honor to be Acknowledged the First Human Being that Scientifically Reconciled Science and Biblical Creation

- Connect with practical tips about how to decode the origin of the universe, life, and chemicals and protect yourself from wrong theories in the literature and the media
- Disrupt all religious and scientific chains of repetitive nonsenses about the universe-origin, life-origin, and turn your attention toward unconventional ideas leading to greater innovation and prosperity
- Empower and align yourself with Science180, the historic company that has done what no other organization has ever done: accurately decode the origin of the cosmos and its content
- Empower yourself to leave unforgettable marks and to stand tall as a symbol of freedom, power, creativity, and originality in your field of expertise
- Enjoy multiple origin-related programs and choose the ones that best suit your needs
- Fearlessly push the boundaries of the human abilities to properly understand what is perceived as un-understandable, mysterious, supernatural, unimaginable, impossible, and unthinkable that holds you back
- Free yourself from boring explanations of the origin of the universe, life, and chemicals and embrace the proven theory that opens doors to unparallel opportunities
- Get inside secrets about how to locate flaws in origin-related theories so you can save time, money, and other resources to improve lives
- Have a reliable access to the world's authority on origin-related matters and get your origin questions professionally answered with the truth step-by-step
- Protect yourself and loved ones by keeping all of you secured and empowered with the true knowledge of the origin of the universe
- Satisfy your burning desire for freedom from beliefs and scientific theories about the universe-origin and life-origin that suffocate you and bind your mind, faith, unbelief, heart, and education
- Scientifically test and know whether there is a God that created the universe or not, and which God it is
- Stand as the lightning bolt that electrifies your colleagues who are still struggling to understand the universe-origin
- Ultimately boost your confidence in detecting, confronting, and avoiding wrong theories by knowing the facts and processes involved in the formation of the universe

To register or to learn more, visit Science180Academy.com today.

Science180: Bringing People Together Through the Power of the Accurate Decoding and Understanding of the Universe-Origin and Life-Origin

3.4. Split-gathering of the fluid layers of the precursors of bodies

Since the beginning of the turbulence that molded the universe, most things in nature have been moving, although some changes may be hard to perceive. Some changes occurred inside the fluids of the precursors of celestial bodies. In *"Turbulent Origin of the Universe"*, I showed that the development of turbulence involved many processes related to the flow, fragmentation, squeezing, and gathering of fluids, and that these processes occurred dynamically during the universe's formation. Likewise, the fluid layers of the precursors did not remain together forever; they had to separate as they took different paths. Here, I will provide some details on the separation and collection of the fluid layers of the precursors of the celestial bodies. Beforehand, I will explain how fluids break up in nature.

Indeed, fluid breakup is the process by which a single fluid mass breaks into several smaller fluid masses. For a fluid to break, it usually goes through an elongation period during which a thin thread-like region forms between larger nodules of fluid. The thread-like region continues to thin until it breaks, and individual daughter fluid bodies are formed. For instance, when water comes out from a leaky faucet, it forms a drop, stretches, and the bottom of the drop is larger than the portion near the faucet where the break will occur. Fluid breakup involves many processes, including the formation of a fluid ligament and neck. In simple terms, a fluid ligament is a mass or a block of fluid connected to a bulk of fluid from which it flows. A fluid neck is a thread of fluid connecting two bodies of fluid about to break up.

Fluid breakup was one of the events that took place during the formation of the universe. Indeed, during the formation of celestial bodies, fluid ligaments were formed as fluids flowed from mother precursors into their daughter bodies. As the fluids of the precursors of the secondary bodies were escaping, the fluids in the precursors of their primary bodies formed a fluid ligament connecting these fluid masses. Just as the neck of a human being connects the head to the rest of the body, a fluid neck connects two bodies of fluids before they break up. Another example is that, just as water leaving a faucet or a nozzle forms a filament (water drip) attached to that faucet or nozzle, and then forms a neck, so also when fluids were leaving a mother precursor to go into its daughter's bodies, a neck was formed. That neck allowed the mother precursor's fluid to flow into its daughter bodies until the ligament pinched off or the neck broke, freeing the daughter bodies. Because fluids are not solid, they do not break sharply like wood or brick; instead, they stretch, reducing the size of the breaking region until a thin layer of fluid is reached, at which point breakup occurs. In summary, for the fluid layers of the precursors to split from one another, fluid ligaments were formed, which thinned until they broke, freeing their daughter bodies on opposite sides of the breaking point. In *"Turbulent Origin of the Universe,"* I explained the process of fluid breakup in detail; if you want to know more, please consult that book.

After the fluid layers of the body precursors separated, they were gathered into their daughter bodies. This gathering occurred not only on the scale of the vortical structures formed within the celestial bodies, but also on the scale of the entire precursor. The turbulence in the fluids caused them to cluster into vortical structures. At the level of the fluid layers, entire layers swirled, collecting all their fluids into one body. The momentum of the fluid flow and the fluid's shear could also have helped accumulate the fluid layers. Just as a fluid parcel can collect into a vortex, so also fluid layers of the precursors of celestial bodies were deformed and amassed into celestial bodies. Because of the interaction between the fluid layers, vortices formed as they flowed, indicating that the layers were being gathered internally even before their split was complete. In other words, as the precursors of bodies were splitting, their constitutive matters were also being gathered together. After the fluid layers split, their gathering continued. Therefore, the splitting and the gathering of the fluid layers cannot be separated from one another as if one finished before the other started. To express the simultaneity of the occurrence of the split and the gathering of fluid layers of precursors, I invented the term "split-gathering together" or "split-gathering" in short. In other words, when I say that the fluid layers of the precursors of a system of bodies were split-gathered, I meant they were split and gathered together into different daughter bodies.

The size of the bodies affected the degree of compactness of their matter. Hence, some bodies are denser than others. In other words, the forces that squeezed the bodies also affected their density. The precursors of the celestial bodies were squeezed into their daughter bodies in different ways. Some were squeezed more tightly than others. For instance, the largest fluids were not compacted tightly. Almost just as water can be in the form of solid (ice), liquid, or gas under different temperatures, so also during the formation of the celestial bodies, the same precursor bodies were molded into different daughter bodies depending on the environment and the process that molded their precursors. In Fig. 11, I illustrated how fluid layers were gathered together into a spherical celestial body (I refrain from using the word "coalesce" because of misconceptions surrounding it in the cosmological literature).

Fig. 11: Gathering together of the layers of fluids in the precursor of a body into a spherical body

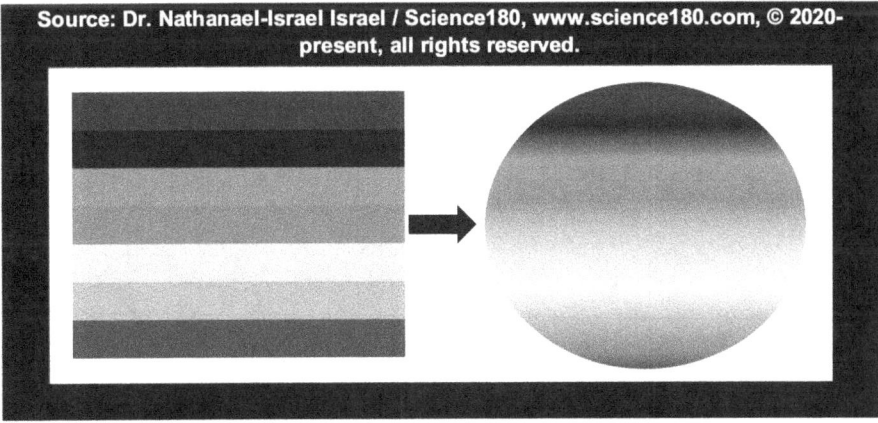

3.5. Why the 99% versus the 1% rule for the system-additive variables

As a reminder, postulate 6 states that during the fragmentation of the precursor of a system of celestial bodies, more than 99% of the mass, kinetic energy, and volume went into the precursor of the primary body, while less than 1% went into the precursor of the secondary bodies. This trend was caused by a natural law (that I discovered) about how the precursor of the celestial bodies split during the genesis of the universe. Indeed, as the precursors of the secondary bodies were escaping the precursor of their primary bodies, a specific amount of fluid flowed into them. To make a long story short, I proved in my scientific book *"Turbulent Origin of the Universe"* that:

- During the split-gathering of the precursor of the Solar System, more than 99% of its mass, kinetic energy, and volume went into the precursor of the Sun, while less than 1% escaped the precursor of the Sun and went into the precursor of the bodies orbiting the Sun.
- Then, during the split-gathering of the precursor of the bodies orbiting the Sun, more than 99% of the mass, kinetic energy, and volume went into the precursors of the planetary systems, and less than 1% went into the precursors of the asteroids.
- During the formation of a planetary system, more than 99% of the mass, kinetic energy, and volume went into the precursor of the primary planet, while less than 1% escaped it and went into the precursors of its satellites.
- During the split-gathering of the satellite precursors, more than 99% of the mass, kinetic energy, and volume were concentrated in satellites formed in the most turbulent zone, usually Zone 3.

After the fluids of the precursors split as mentioned above, they were gathered together into celestial bodies.

3.6. Cause of the variation of the eccentricity and orbital inclination

In postulate 9, I showed that in planetary systems, most satellites with the highest eccentricities, highest orbital inclinations, and smallest radii are usually located farther from the largest satellites. This trend is because the way the fluid layers of the precursors interacted with one another deformed the precursors of their daughter bodies. For instance, as the precursors of some satellites were being positioned by the process that gave rise to them, something must have acted on them at some point to tilt their orbits and elongate some of them. Considering postulate 10—which stated that, when a fluid column is squashed (meaning compressed, squeezed, or pressed by others above it), its rotation slows down, and the reverse happens if the fluid column is stretched—I deduced that, the stacking of the fluid layers in the precursors of the satellites increased the rotation of the structures formed in the squashed layers. But, once the fluids of the precursors of the largest bodies—which were squashing layers beneath them—split from the fluids of the precursors of the bodies located below them, the latter continued their stretching as the pressure that was applied to them was released. In return, this stretching elongated the orbits of the bodies located beneath the precursor of the largest ones. At the same time, as I scientifically showed in my book *"Turbulent Origin of the Universe"*, in reaction to the squashing by the precursors of the largest bodies, the precursors of the bodies located beneath the precursors of the largest ones were overturned or flipped over, therefore tilting the orbital plane of the affected bodies. Hence, in most planetary systems, beyond the largest satellite, the eccentricity and the orbital inclination are usually high. But before the largest satellites, the eccentricity and orbital inclination are minimal because the fluid layers of the precursors of these bodies were not squashed by the fluid layers of the largest bodies.

Just as when two people holding hands begin to pull apart, they can be pushed in different directions when their hands are suddenly released, so also during the breakup of two consecutive precursors, one could have been pulled forward and the other backward. The presence of prograde satellites (less tilted) among retrograde satellites (very tilted) can be linked to how the orbits of their precursors were flipped or tilted after their precursors broke from one another and were then pulled or pushed in different directions. Imagine 20 people or more holding hands one to the other (as if in a human chain), with ten pulling in one direction and the other 10 pulling in another direction. According to the pulling strength of the people at the time of the breakup or separation of their hands from one another, they would not all fall in the same direction, nor with the same strength. Some would fall forward, while others would fall backward, and some may not even fall at all. Some may stumble to the left or right to regain their balance. The same thing happened for the precursors of celestial bodies: their fluid layers were connected before separating, yet exerted some pull and push on one another; at the time of their breakup, they were thrown in different directions depending on the prevailing conditions and their location relative to the point of breakup.

This process could have also affected the axial tilt of the bodies. During the breakup of the precursors from one another, the rotational axis of some bodies was pushed forward or backward according to the direction of the fluid flow that affected them. For instance, as the fluids moved, a strain was also built up in them. At one point, the rotational axis of the celestial bodies was bent. The celestial bodies with the highest axial tilts are for their system what a knee is for the leg and what an elbow is for the arm. They are like the hinge around which the fluid ligament of the precursor of the bodies downstream of them was pushed and tilted toward the edge of their system. That is why outside the celestial bodies with the highest axial tilt, the orbital inclination of the bodies is usually high.

3.7. Explanation of the density of the bodies

As I was working on the density of celestial bodies in 2017, I was inspired to ponder how some houses are built in rural parts of the world where I grew up. Here, I will share with you an experience I had while growing up about building houses with mud balls made from soil, sand, and water, and how it can be applied to the formation of the universe. This experience reminded me that construction techniques can help explain the shapes and densities of some celestial bodies. Indeed, as I was growing up, people in my town used to build houses by first digging the ground, preparing a foundation, collecting soil, adding water to make mud, and then erecting the building. The mud was usually rolled into balls by hand and then collected into spherical bodies about the size of a soccer ball. Then, those balls were carried to the building site and laid one on top of the others according to the building's design.

To make these balls, the builders used their hands to apply a force to a quantity of mud matching their age and strength. People who were strong gathered large balls of mud, while those who were weak or young gathered small balls of mud. Depending on how their hands applied a force to the mud being rolled on the ground, some balls ended up being a perfect sphere, others were oblate, others were prolate, and others did not even have a defined shape. The density of the balls depended on the compression or squeezing applied to them as the mud was rolled. The balls were turned over and rolled in various directions (e.g., north to south, east to west) to ensure their shapes were nearly spherical. As the builders were making and shaping the balls, they were simultaneously compressing and squeezing them. In the end, the size of the balls and the compression they received were generally inversely proportional: the bigger the mud ball, the harder it was to apply a strong force to the resulting ball. But the smaller the mud ball, the easier it was to compress it and roll it on the ground to make a denser ball. After the mud balls were made, the builders stacked them one on top of the other, pressing them to build the house. During the process, layers of the mud balls were laid and left to dry so the buildings' walls would not collapse due to their weight and moisture. People who are rich use bricks made with concrete or cement, which not only stand longer and are stronger but also better resist other challenges, such as rain, which can bring the building back to its original materials: soil and water. Fig. 12 illustrates part of what I said above.

Nathanael-Israel Israel: Has had the Honor to be Acknowledged the First Human Being that Scientifically Reconciled Science and Biblical Creation

Fig. 12: Building of a house with the mud and bricks made with soil and water

By recalling this experience that I had when I was growing up, I felt like a similar process could have been used to make the celestial bodies:

- Create chemical particles
- Mix the particles to make complex ones
- Mix the particles and compounds with fluids (e.g., water) to lay some foundations
- Compress the bodies

As of 2021, I perceived that the strata found on some planets, like Earth, are like the bricks a builder lays while erecting a house. A foundation was laid for the Earth, and its mud was used to make part of the Earth's crust. Driving through roads with cliffs allowed me to easily see how layers of materials deposited like brick fill the Earth's crust.

As a take-home message regarding "why celestial bodies have different densities and why some are denser than others," I would say that, as the fluid layers of the precursors of the bodies in the Solar System started collecting together to shape their bodies, their fluids were spiraled and wrapped together as vortexes of various sizes were formed inside of them; a vortex being a region in a fluid in which the flow is rotating around an axis line. The speed of the fluid layers affected the way the spirals squeezed their daughter bodies that were being formed, but the ability of these daughter bodies to rotate also affected the intensity with which the daughter bodies were squeezed. The higher the speed of the fluids in the layers, the higher the likelihood that their daughter bodies would be more squeezed. But when the daughter bodies could not rotate faster, their ability to squeeze their contents and become denser decreased. Consequently, the densest bodies are those for which the orbital speed and the rotational speed are balanced and relatively high. The higher orbital speed of the precursors, combined with the ability of their daughter bodies to rotate faster, could have affected the daughter bodies' density. That is why, although the terrestrial planets are the densest, the Earth is denser than all of them because it has a relatively higher balance between its orbital speed and rotational speed. In contrast, although Mercury and Venus have higher orbital speeds, their densities are lower than Earth's because their precursors were unable to rotate quickly, completing a single rotation in months, whereas the Earth completes a full rotation every day. Similarly, although the giant planets have a higher rotational speed, their densities are not the highest.

The low density of the Saturnian planetary system's precursor may also explain the large size of the Saturnian rings. Because the precursor of the Saturnian planetary system was unable to gather together all its matter into the Saturnian satellites, the matter that was not incorporated into the satellites formed the rings. I will provide more details about the rings later.

3.8. Why some planets giant gas, ice, and terrestrial

As the precursors of the celestial bodies were separating from one another, the particles inside of them were reorganized according to their size, position, and energy. Some of the fluid chunks were too large for the conditions around them to make them stick together into a solid. Hence, some precursors yielded gas-filled bodies, others formed icy bodies, and others formed solid bodies like concrete. For instance, Jupiter and Saturn are giant gas planets because their precursors were unable to coalesce into a solid, hard body like the precursors of the terrestrial planets (e.g., Earth) did. Hence, Jupiter and Saturn are called giant gas. Likewise, the precursors of Uranus and Neptune yielded 2 giant icy planets; hence, Uranus and Neptune are called giant ice. Inside the giant gas, the fluid at the surface and/or in the atmosphere was organized to this day. until today. And these cyclones are a reminder of the turbulence that prevailed during the universe's formation. Because conditions allowed the Earth's precursor to solidify at its surface, the Earth has a hard crust. The extent of the squeezing, compaction, or compression of the vortexes of the precursors of the celestial bodies affected the density and many other parameters of the celestial bodies. Bodies that were powerfully or tightly squeezed are denser than those lightly squeezed. Because the vortices of larger bodies were loosely squeezed, their density is generally small, and their surface is not always crusty. That is why the Sun, for instance, is very large, yet less dense and denuded of a crust at its surface.

3.9. Semi-major axis and expansion of the universe

It took some time for the fluids in the layers could be gathered together to form celestial bodies. For instance, after escaping the precursor of its primary body, the precursor of the secondary bodies flowed for some time before splitting into daughter bodies separated by specific distances. The distance that the fluids of the precursor of a secondary body traveled (after escaping the precursor of its primary body) before being gathered together into a celestial body is about its semi-major axis. In each system of bodies, the distance between the celestial bodies was defined by the duration of the process that amassed the fluids of their precursors. The longer it took before some fluids could be amassed to form a body, the farther they could have traveled. Fluids that escaped their mother precursors slowly may not have traveled very far before being formed. For they may not have had enough energy to move for a long time.

The celestial bodies were positioned at different distances according to the conditions prevailing during their formation, including the kinetic energy of their precursors and their positions within the stack of fluid layers they once belonged to. For instance, because the fluid layers of the precursors that were on top of the stack split first and the fluid layers at the bottom split last, the semi-major axis of the celestial bodies increases from the innermost bodies to the outermost bodies. The viscosities of the precursors of the celestial bodies could also have affected the distances separating their daughter bodies. For instance, a very viscous fluid could have stretched over a long distance before breaking into pieces. How the fluid layers broke from one another affected the outcome of their daughter's bodies.

Using the semi-major axis, I calculated the semi-major axis increment, which is the distance separating a body from another one located either before or after it. I found that the semi-major axis increments of the bodies that I studied in the Solar System ranged from zero to millions of kilometers. In other words, while some bodies shared almost the same orbit, others are millions of kilometers away from their neighbors. Before I close this section, I would also like to mention that, as of today, scientists believe the universe is expanding. When I looked at the data I analyzed, I felt that the expansion of the universe can be traced back to the explosive, mysterious scattering of the bulk of the universe's original matter at the beginning of the universe's formation. Since the beginning, the natural tendency of the distance separating celestial bodies in the universe is to expand. This does not mean that all bodies in the universe are getting larger, but rather that the distance separating them is increasing, as if the universe were being stretched.

3.10. Take-home message

The take-home is that the orbital speed and the rotational angular speed of the celestial bodies were determined by how the fluid layers of their precursors interacted with one another; ultimately, some layers squashed others, which were then stretched, elongated, overturned, pushed, or pulled away. I showed that the orbital inclination, eccentricity, rotational speed, size, and many other characteristics of celestial bodies are not caused by random events or mere chance, but were calibrated by the turbulence that prevailed during the universe's formation. The rotation and revolution that some bodies display today were progressively born out of the turbulence that took place during the formation of the universe.

'Science180 Academy' Success Strategy:
SCIENCE180 SERVICES AND PRODUCTS YOU WILL LOVE

Because you are reading this book, you are probably very interested in answering your questions about the origin of the universe, of life, and of chemicals. Imagine you want to be trained by Dr. Nathanael-Israel Israel and his team so you can benefit from their outstanding expertise to empower yourself or your team. Or you want him to give a keynote speech, a seminar, or any other kind of talk or conference at your organization. Or you want him to mentor you or some people or team at your organization. Maybe you have critical origin-related questions that you need his help to accurately answer. You want a true expert to talk with you about the customized program or game plan that fits your needs. You want him to tailor his advice, expert feedback, and proven shortcuts to the stage of life you are in and help you get to where you want to be in your desire to properly understand the origin of the universe, life, and chemicals and harness the benefits that come with it. Perhaps you don't know how to properly get any of these important tasks done according to your specific needs or the needs and demands of your organization. That is what Science180 Academy is all about. Visit Science180.com/services for more details about how to benefit from the services that Science180 provides.

Maybe you are a leader that wants to hire Dr. Nathanael-Israel Israel and his team to train some departments at your organization. Or you want to refer them to other companies like a good dish passed around the dinner table, and you want to explore how Nathanael-Israel Israel can pay you something for that referral. Maybe you attended Nathanael-Israel Israel's speaking program, for which, without going into details, he accurately raised your awareness about how the universe, life, and chemicals were formed.

Nathanael-Israel Israel: Has had the Honor to be Acknowledged the First Human Being that Scientifically Reconciled Science and Biblical Creation

Or maybe you attended his training, in which he detailed and showed you how he decoded the scientific data using various tools and certain thinking strategies that helped him and which transferred some skills to you; and now, you are interested in a long term one-on-one consulting, or mentoring program with him, so that, he delves into more details about how to use proven techniques to decode the universe (strategies for data collection, data analysis, data presentation, writing, and even tips for future research) and change your behavior on a long term basis. If you related to any of the points mentioned above, Science180 Academy is the right fit for you! Other customizable services that Science180 provides include: Assessments, Books, Book publishing (Yes! Science180 can publish your books (learn more at www.Science180Publishing), Conferences, Consulting, Executive mastermind groups, Master classes, Online courses, Podcasting, Seminars, Speaking engagements, Survey and research tools, Training, Video programs.

Here are other reasons why you should choose to work with or hire Nathanael-Israel Israel and the team at Science180:

- A simple universe-origin, chemicals-origin, and life-origin theory that made no assumption
- Accurate universe-origin, life-origin, and chemicals-origin decoding trailblazer
- All the products and services you need to accurately and easily decode the universe-origin, life-origin, and chemicals-origin
- Complex universe-origin, life-origin, and chemicals-origin questions solved accurately in a simple language
- Customizable universe-origin, chemicals-origin, and life-origin trainings with unique materials
- Easily understand complex universe-origin, chemicals-origin, and life-origin equations in minutes
- Efficient, trustworthy, and cost-effective company to add to your strategic journey toward your best tomorrow
- Fearless universe-origin, chemicals-origin, and life-origin decryption trailblazer
- Light in the heart of science the lamp of understanding
- Nonconformist, rule-breaker, and accurate demonstrator of the universe-origin, chemicals-origin, and life-origin
- Personalized universe-origin, chemicals-origin, and life-origin decoding package
- Source of unconventional wisdom and knowledge on the origin of the universe, life, and chemicals
- Learn more at www.Science180.com/services

Nathanael-Israel Israel: Acknowledged as Undisputable Specialist of all
Questions at the Intersection of Science and Biblical Creation

CHAPTER 4

IS EXPLAINING THE UNIVERSE BEGINNING THIS WAY GOING TO SET THE STAGE FOR THE SCIENTIFIC REVOLUTION AND ENLIGHTENMENT OF THE 21ST CENTURY?

4.1. What was it like to be the first matter in the universe?

The universe has a beginning, which, in light of the scientific evidence I studied, I will present to you in this chapter from a new perspective. The explanation of the origin of the universe is challenging, for no human being was present to recount the story, but that story can be retold using facts abounding in the universe. Although scientific and religious books have addressed the question, many details surrounding it remain. For instance, while most Christians believe that *"In the beginning, God created the …"*, they struggle to scientifically demonstrate how God created the universe using His words revealed in the Bible. Likewise, while most nonbelievers would say that *"A few seconds after the beginning, there was a Big Bang …"*, they also struggle to prove where the first particle came from and what was the driving force behind the giant explosion claimed by the Big Bang. In this chapter, I do not pretend to have all the answers, nor do I intend to argue about these statements, but based on my careful study of the characteristics of the universe, I have unearthed key details about the beginning that I would like to bring to your attention.

In the beginning of the universe, a certain kind of matter, which I called the "turbulent prima materia," mysteriously appeared in the universe out of "nothing" …, and through very complex, dynamic, and turbulent processes, it was progressively shaped into all types of bodies known in the universe today. The turbulent prima materia was the mother of all bodies found in the universe. As of today, we are surrounded by many complex things and beings, some of which began as simple entities but developed into complex structures, which can be understood by studying the laws and processes behind their formation. For instance, as of today, most living things start with a cell that first divides into undifferentiated cells (capable of

49

becoming any kind of cell), which then go through stages of differentiation and split into different parts with different functions.

In each seed, for instance, there is a program that allows it to germinate and grow into a mature organism. Likewise, the universe began with original matter, which had the potential to become any type of matter. After splitting and undergoing changes, this original matter gave rise to daughter bodies that became the precursors of the celestial bodies and other matter in the universe. In other words, just as during the development of a living organism, an egg can be differentiated into different types of cells, tissues, organs, apparatuses, or systems, so also during the formation of the universe, the "turbulent prima materia" (i.e.,, the initial matter in the universe was programmed to yield different types of matter according to the conditions that subdued it. While it is possible and easy to observe a living thing growing from the stage of an egg to the stage of an entire adult, and that scientists have gathered extensive details on how life developed, it is not possible to witness the process that birthed the celestial bodies (for that process was completed and no new celestial body is being formed now as it was in the beginning). But by carefully studying the characteristics of celestial bodies today, I was able to pinpoint key details about the machinery that shaped the destiny of the turbulent prima materia (the original matter in the universe) and distributed it across various scales.

In my well-acclaimed book *Turbulent Origin of the Universe*," I extensively talked about the characteristics of the turbulent prima materia, but here I will introduce just some of them. Indeed, the initial matter in the universe occupied a very huge, wide, and deep portion of space. It was formless, and the processes that shaped it are responsible for the forms of matter seen in the universe today. Hence, I called the turbulent prima materia the "mother" of all bodies in the universe. Although it sounds easy to label the initial matter as having a specific form or as being made of some chemical elements known today, the work I have done suggested that giving a shape or form to the initial matter and then trying to use it to explain other matters in the universe is one of the reasons some previous scientific theories have failed to explain the formation of the universe. Considering the initial matter as shapeless also placed a greater burden on me to clearly explain how it was converted into the current forms of matter and bodies in the universe.

Matter exists in a few forms. Three states of matter are generally taught in elementary schools: solid, liquid, and gas. However, at an advanced level, experts in the field added another state called "plasma," which is none of the three I just mentioned. For instance, the Sun is believed to be made of plasma. By the way, the term "plasma" I just referred to isn't the same as the plasma in your blood. I showed in my other books that the state of the "turbulent prima materia" was none of the states of matter known today. For the "turbulent prima materia" to go through complex changes and developments before each of the current states of matter could be formed. Because every particle in the early universe was changed by the processes that fashioned the world, nothing in the world today is a precise duplicate of the initial matter at the beginning. However, I felt like it did not take too long before the state of the "turbulent prima materia" could have been near that of a plasma or fluid. The original matter at the beginning of the universe could have looked like a flame, a burning fire, or something very hot like magma or lava. The plasma in the Sun and the magma or lava that erupt from Earth's interior are witnesses to what the initial matter could have been made of.

Soon after its creation, the initial matter in the universe was scattered, as I termed the original mysterious scattering. This can also be called a sudden burst asunder, a breaking open, or a breaking apart of the initial matter in the universe. In other words, the bulk of the turbulent prima materia was violently broken open or broken apart by a noisy process accompanied by a series of "explosions." This major explosive event marked the beginning of the spread of the precursors of bodies in the universe. Due to the movements that followed the initial scattering, the place that celestial bodies are today is not where their precursors were in the beginning. In other words, the bodies' current positions in the universe are not where their existence and journeys began. The precursors of many celestial bodies were relocated as the particles within them were shaped and moved. The original mysterious scattering split the turbulent prima materia into chunks of matter, which were propelled in many directions. The original mysterious scattering was not merely an explosion but an event that imparted specific motions to chunks of the turbulent prima materia, driving them into major turbulence. In the famous book called *Turbulent Origin of the Universe*," I explained the factors that contributed to the breaking apart of the initial matter.

During the original mysterious scattering, chunks of matter of various sizes were formed. Some of the astronomical chunks of matter became the precursors of galaxies. At the microscopic level, small amounts of matter were collected, forming the foundation of atoms and their constituents. Cascades of "explosions" or breakups occurred on various scales until a point when no further breaking apart could occur. In other words, as the clusters of matter were distancing themselves from the position of their mothers, a series of "explosions" followed as daughter "explosions" were born out of mother explosions and so on and so forth until the remaining clusters could no longer "explode." Due to the decrease in size and pressure of some daughter bodies along the sequential cascades of breakups, their explosive ability declined. Consequently, the clusters of precursors of bodies stopped "exploding" at one point, ending the chain of "explosions" that followed the original mysterious scattering. Meanwhile, changes were occurring in the clusters of matter as they were broken apart and assembled. In *Turbulent Origin of the Universe.*" I extensively showed that the chunks of matter were changed by complex events comprising their fragmentation, squeezing, and others, which "destabilized" and "reshaped" them:

- Transfer of energy
- Initiation of movement (e.g., revolution and rotation)
- Flow and mixing of flow
- Formation of fluid layers
- Stretching, tilting, overturning, and squeezing of fluids and structures formed in them

- Sizing, positioning, and spacing of bodies
- Gathering together of fluid layers
- Spiraling or spinning of fluid layers
- Birth and strengthening of various forces, etc.

In this book, I will explain how the events I listed above shaped the destiny of the turbulent prima materia, yielding the bodies known in the universe today. These events did not necessarily occur in the order I listed them. The bulk of the turbulent prima materia underwent fragmentation, simultaneously accompanied by modifications in the characteristics of their daughter bodies.

Just as the precursors of the bodies in the universe began moving, time as we know it was born. Before that moment, there was no physical thing (as ordinarily known today) in this world yet that could have been used to determine time. From a religious viewpoint, although I personally believe that an eternity existed before the beginning of this world, I will not delve into those details here, but in another book. Due to the energetic nature of the initial matter in the universe, the bodies in the universe are filled with energy. Part of the energy communicated to the precursors of bodies was used to mold and move their daughter bodies. Another part of the energy that was in the precursors of the bodies was used to tilt, incline, or flip over the orbital plane and rotational axis of some bodies. Part of it imparted speed to the bodies, and part was stored in the particles constituting these bodies, according to their size. Because the original matter in the universe was moving when its daughter bodies were being formed, every matter or body in the universe contains particles that are moving. Likewise, although we cannot see it with the naked eye, and some people may not believe it, everything in the universe is moving. I remember driving and talking to someone about everything moving in the universe, and that person didn't understand or accept it. I guess that person had no idea where that conversation came from and where it was going. *"Ten years later, here we are with a lot of books written about the formation of things in the universe,"* that person confessed to me a few months ago.

4.2. How was turbulence (known as the unsolved problem in science) in the universe born?

One of the key things that occurred during the formation of the universe is turbulence, a complicated phenomenon considered as *"the last great unsolved problem of classical physics."* Although efforts to understand turbulence have persisted for more than 500 years, many questions remain. I have been talking about turbulence, but here, I will provide a few more details about its origin and development.

Turbulence can be seen in the movement of cream poured into coffee or the movement of clouds in the atmosphere. Turbulence is usually considered as the state of a moving fluid characterized by a "random," "disordered," "confused," and "chaotic" multidimensional vorticity (vorticity being a property related to how fluids swirl or wrap around like vortexes or eddies). However, during my research on the origin of the universe, I realized that turbulence can lead to an ordered state of matter that scientists are unable to comprehend because they have been approaching it with a limited mindset, unable to know that every matter in the universe is a product of the original turbulence at the beginning of the universe. Turbulence contains too many details, and grasping the full picture of its information is impossible for those who like to think in straight lines. I have established that the turbulence that took place during the formation of the universe is the "mother of all turbulences," for it birthed everything in the universe, and no other turbulence is bigger than it.

But what caused the initial turbulence in the universe? Indeed, it has been proven in laboratory settings that, when fluid layers move past one another at different speeds, instability can start at their interface and grow into turbulence. I already showed that, during the formation of the celestial bodies, fluid layers formed and, because they were stacked on top of one another, moved at different speeds, with the fastest on top and the slowest at the bottom. Therefore, due to the size and the varying speed of the fluid layers of the precursors of the celestial bodies, turbulence occurred in them. Because vortices are commonly formed in turbulent fluids, the turbulence that developed in the precursors of the celestial bodies also led to the formation and rearrangement of matter as vortices. The intensity of the turbulence was not always the same from one layer to another and from one precursor of matter to another. Not only did turbulence contribute to splitting the bodies' precursors, but it also imparted distinct characteristics to them depending on their location. I understand that the movement that celestial bodies exhibit today can be traced back to the major instability that occurred during the formation of the universe.

The fluids in the turbulence at the origin of the universe had structures within them on all imaginable scales, from the invisible (or the tiniest imaginable) to the largest conceivable. The bodies at each turbulence scale went through different developmental phases and gave rise to various forms. The energy that the daughter bodies received from their mothers affected their instability and turbulence. The kinetic energy that some precursors received from their mothers allowed them to overcome the eventual resistance posed by the viscous forces of their constituents. Here, look at viscosity, like what makes honey or molasses sticky, while water is not sticky. In other words, viscosity is the physical property that causes honey to drip slowly but water to flow very easily. Because of factors such as viscosity, the fluids in the precursors of the bodies could have been subject to varying degrees of turbulence, which also led to the formation of bodies with diverse characteristics. Although the objective of this book does not allow me to detail them, the scales of the turbulence that I studied in my other books included those of the:

- Invisible or spiritual scale that no naked eye or scientific equipment will ever see
- Subatomic scale: scale of the currently known subatomic particles
- Atomic scale
- Molecular scale
- Mineral scale

- Rock scale
- Asteroid system scale
- Satellite scale
- Planetary system scale
- Stellar (related to star) system scale
- Galactic (related to galaxies) scale
- Clusters of galaxies scale
- Scale of the whole universe

4.3. Origin of the precursors of the bodies in the universe

Each celestial body in the universe today had a precursor, meaning a body that came before it and from which it was formed. Knowing the precursors of the celestial bodies is important for understanding the process that gave rise to the universe. Here, I will help you understand the origins of the precursors of all bodies in the universe. To achieve that goal, I will build on what I already demonstrated. Indeed, I showed that the original mysterious scattering dispersed the bulk of the original matter into clusters that started moving. I also showed that fluid layers formed in the precursors of the celestial bodies and flowed in specific directions, depending on how their mothers were split. For instance, I demonstrated in the previous chapters that the precursor of the Solar System first split to yield the precursor of the Sun and the precursor of the bodies orbiting the Sun. Then, the precursor of the bodies orbiting the Sun flowed away from the precursor of the Sun and were split-gathered into various precursors, including the precursors of the planetary systems, which, in turn, split-gathered into the precursors of the planets and the precursor of the satellites. The organization and movement of the celestial bodies suggested to me that their precursors were formed during a sequential cascade breakup during which a mother precursor broke up and was reorganized into daughter bodies, which, in their turn, did the same and so on and so forth, until no breakup could occur (see postulates 1 and 2). In other words, just as, in some families, some parents have children, grandchildren, and great-grandchildren, so also during the formation of the universe, the precursors of some systems of bodies split and gathered into other precursors in a cascade that could produce sets of daughter bodies, which could be called daughter bodies, granddaughter bodies, and great-granddaughter bodies of the mother precursors. Because the universe is organized as a nest of primary and secondary bodies, the precursor of the systems of bodies could have birthed both the precursors of the primary bodies and the precursors of their secondary bodies.

To download a pretty and colorful version of the graph illustrating the cascade of breakup of mother precursors into daughter bodies and so on and so forth, visit www.Science180.com/TurbulentPrimaMaterialBreakup

Unlike what most people think, celestial bodies did not descend from one another but from specific precursors. For instance, the planets did not descend from the Sun but from the precursor of the Solar System. Likewise, the satellites did not descend from their primary planets but from the precursor of their planetary system. Therefore, it is incorrect to model satellites after the characteristics of their primary planet as if the planet birthed the satellites. For instance, imagine a family consisting of siblings of various sizes and ages. It can be misleading to model the siblings in that family by treating the younger or smaller ones as dependents of the older or larger ones. Although they can influence one another, younger or smaller siblings do not depend on, nor were they born by, older ones, and vice versa. However, all the siblings (the older or the larger siblings and the younger or smaller ones) come from their parents. In other words, the best way to model the origin of siblings is to refer to their parents. Although some parents may be dead, by carefully explaining the characteristics of the siblings, it may be possible to explain what the parents may have looked like before their death. Similarly, although the precursors of the planets (considered here as the bigger siblings) and the precursors of their satellites (considered here as the smaller siblings) are dead (for they no longer exist), it is possible to properly estimate the characteristics of these precursors by carefully examining the properties of the planets and satellites and by considering them as descendants of the precursor of their planetary system. Likewise, the planets in the Solar System should not be considered to descend from the Sun, but rather from the precursor of the Solar System (which includes the precursor of the Sun, the planets, and the asteroids). As a rule, it is misleading to model secondary bodies after the characteristics of their primary bodies. In a specific system, both the primary body and its secondary bodies descended from the same precursor: the precursor of their system.

Seeing the organization, features, and the existing classification of celestial bodies in the universe, I showed in *"Turbulent Origin of the Universe"* that the way the precursors were split-gathered can be categorized into 9 groups, which I termed "generations of split-gathering." To save time, I will spare you from the lengthy demonstration I did concerning these generations of split-gathering, but I will mention just a few things. Indeed, I showed that the first generation of split-gathering was the split of the bulk of the original matter in the universe into the precursors of the largest clusters of galaxies, which at their turn were split-gathered into smaller clusters of galaxies and so on and so forth until the precursors of the smallest galaxies were reached and could no longer split-gather into the precursor of a galaxy. Because of this cascade of breakups, the galaxies in the universe are organized into clusters of galaxies, which at their turn are arranged into superclusters, which at their turn are arranged into clusters of superclusters, which at their turn can be structured into clusters of clusters of superclusters, and so on and so forth until all higher levels of clusters of galaxies are included in the largest system called the universe. The precursor to the Milky Way Galaxy (the galaxy the Sun belongs to) formed around this stage in the universe's formation. In fact, most of (if not all) the stars we see with the naked eye in the night sky belong to the Milky Way.

As the precursors of the galaxies were split-gathering, precursors of stellar systems were formed. At this stage, the precursors of each galaxy split-gathered into the precursors of several stellar systems. It was around this time that, as its fluids were split-gathering, the precursor of the Milky Way Galaxy gave birth to the precursor of the Sun. As the cascade of split-gathering continued, the precursors of the stellar systems in the universe birthed the precursors of their stars and the precursors of the bodies orbiting these stars. In other words, by this stage in the genesis of stellar systems across the universe, precursors of stars and of the bodies orbiting them were popping up everywhere. For instance, at the moment of the Solar System's genesis, the precursor of the Solar System gave rise to the precursor of the Sun and the precursor of all the bodies orbiting the Sun. As a reminder, in a previous chapter, I showed that the precursor of the bodies orbiting the Sun escaped the precursor of the Sun at about the Sun's escape velocity. After the precursor of the bodies orbiting a star escaped the precursor of that star, it flowed away from the precursor of the star and was then split-gathered. Likewise, at this stage, the precursors of the bodies orbiting the Sun started flowing and split-gathering into the precursors of the planetary systems, the precursors of asteroid systems, and the precursors of all other matters found in the Solar System. In their turn, the precursors of the planetary systems flowed and split into the precursors of their primary planets, satellites, and rings. The precursors of the planets and planetary systems in the Solar System were:

1. Precursor of Mercury
2. Precursor of Venus
3. Precursor of the Earth-Moon system
4. Precursor of the Martian planetary system
5. Precursor of the Jovian planetary system
6. Precursor of the Saturnian planetary system
7. Precursor of the Uranian planetary system
8. Precursor of the Neptunian planetary system
9. Precursor of the Plutonian planetary system

As I already showed, the precursors of the bodies orbiting the planets escaped the precursors of the planets at about the planets' escape velocity. After that escape, the precursors of the bodies orbiting the planets flowed and split into the precursors of the satellites and rings (in some cases). Likewise, the precursors of asteroid systems were divided into those of asteroids and those of their satellites. By this stage in the formation of the Solar System, the precursor of the Sun was gathering into the Sun, while precursors of planets, asteroids, and satellites were forming and accumulating into those celestial bodies. As the precursor of what could have been a planetary system was split-gathering, the conditions (e.g., high viscosity, small size, and low energy) were not met for it to form a satellite around a planet; therefore, only a planet was formed. In my book "Turbulent Origin of the Universe," I logically explained that, as a result, some planets and asteroids lack satellites, while others have many. In general, the split-gathering of the precursors of the secondary bodies in a planetary system was about their split into the precursors of individual satellites and rings. For that to happen, the precursors of the satellite systems underwent turbulence that divided them into precursors of individual satellites and rings. Because the fluids of

Nathanael-Israel Israel: Acknowledged as Undisputable Specialist of all Questions at the Intersection of Science and Biblical Creation the precursor bodies orbiting the Sun were splitting as they were flowing, all the planetary systems and asteroid systems were not formed at the same time. Upcoming is a sketch of the precursors of the planets and planetary system in the Solar System (Fig. 13).

Fig. 13: Layers of fluids in the precursors of the planetary systems in the Solar System

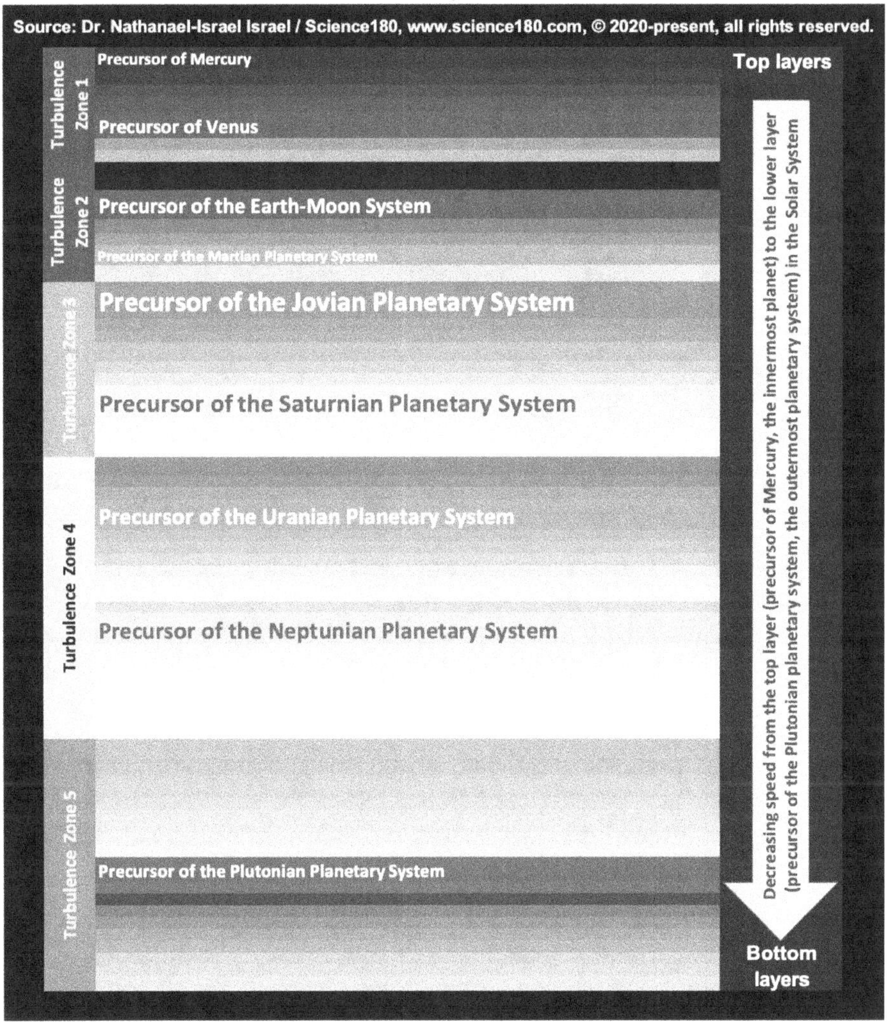

I will now provide some details on how the precursor to the Earth-Moon system was formed (Fig. 14). I already showed that, in a stack of fluid layers, the top layers split first to form the innermost bodies, while the bottom layers split last to form the outermost bodies. Likewise, as the precursor of the Solar System was giving birth to

Science180: Complex Universe-Origin, Life-Origin, and Chemicals-Origin
Questions Solved Accurately in a Simple Language

its daughter bodies, the precursor of Mercury (the innermost planet in the Solar System) was among the first to be born. After the precursor of Mercury split from the stack of fluid of the precursor of the bodies orbiting the Sun, the precursor of Venus split. Then came the turn of the precursor of the Earth-Moon system to split from the remainder of the precursor of the bodies orbiting the Sun. Afterwards, the precursor of the Earth-Moon system split into the precursors of the Earth and the Moon, as exemplified below.

Fig. 14: Precursor of the Earth-Moon System

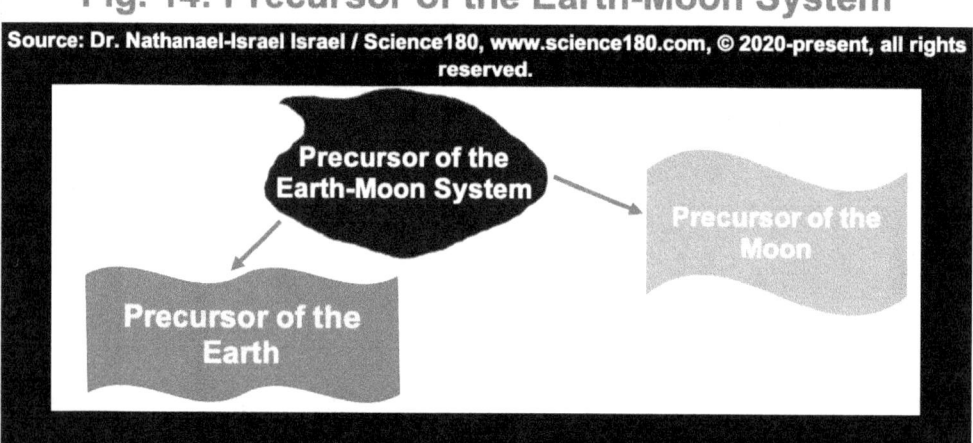

While the precursor of the Earth was being gathered into the Earth, the precursor of the Moon traveled for a while before being gathered together into the Moon. In the next chapter, I will explain the exact distance that the precursor of the Earth and the precursor of the Moon traveled before being formed.

As the precursor of the Earth-Moon system was forming, the Earth and the Moon, the remainder of the fluid layers of the precursor of the bodies orbiting the Sun continued its journey until reaching a point when the precursor of the Martian planetary system split from it and was then split-gathered into the precursor of Mars and the precursor of the Martian satellites. Later, the precursor of the Jovian planetary system split from the remainder of the stack of fluid layers and then split-gathered into the precursor of Jupiter and the precursor of the Jovian satellites and rings. After splitting from the precursor of Jupiter, the fluid layers of the precursor of the Jovian satellites went through another period of turbulence before yielding their satellites. Likewise, the precursor of the Saturnian planetary system split-gathered into the precursor of Saturn and the precursor of the Saturnian satellites and rings (later, I will revisit how Saturn's rings were formed). In its turn, the precursor of the Uranian planetary system was split into the precursor of Uranus and the precursor of the Uranian satellites and rings. However, as it was splitting from its neighbors, the precursor of the Uranian planetary system was tilted or flipped over, causing Uranus and its satellites and rings to orbit the Sun as if they were rolling. The precursor of the

Nathanael-Israel Israel: Acknowledged as Undisputable Specialist of all
Questions at the Intersection of Science and Biblical Creation

Neptunian planetary system was split into the precursor of Neptune and the precursor of the Neptunian satellites and rings. Finally, the precursor of the Plutonian planetary system was split into the precursor of Pluto and the precursor of the Plutonian satellites. After the precursors of all the planetary systems in the Solar System split from the stack of fluids of the precursor of the bodies orbiting the Sun, the rest of the fluid layers of the precursor of the bodies orbiting the Sun continued their journey until the precursors of all the outermost asteroids (located at the bottom fluid layer) were reached. Fig. 15 is a sketch of the precursor of a planetary system, split-gathering into a planet and its satellites.

Fig. 15: Layout (as of today) of the primary planet and satellites born from the precursor of a planetary system

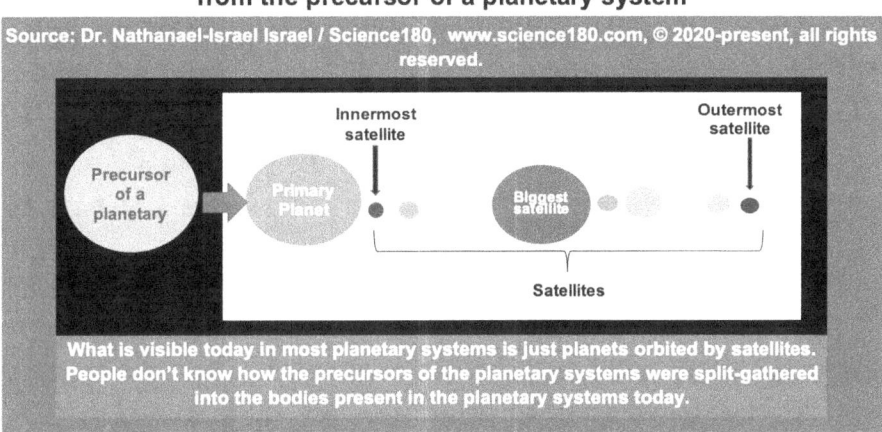

What is visible today in most planetary systems is just planets orbited by satellites. People don't know how the precursors of the planetary systems were split-gathered into the bodies present in the planetary systems today.

As the precursors of the stars, planets, asteroids, and satellites were forming, the matter within them was rearranged, and this internal clustering led to the formation of the precursors of minerals and rocks. For instance, as the Earth was forming, precursors of rocks formed and later gave rise to specific rocks and minerals. Inside the precursor of those rocks and minerals, precursors of atoms, molecules, and chemical compounds were formed. By the way, atoms are what ordinary matter is mostly made of. Some atoms are called exotic atoms because they are different from the ordinary or conventional atoms found on Earth. In other words, in some environments far away from the Solar System, some strange atoms can be found. Inside the precursors of the atoms, precursors of subatomic particles (i.e., particles smaller than atoms) were formed. Just like all other bodies or particles in the universe, the subatomic particles also had their precursors. As I have explained in other books (e.g., "*Turbulent Origin of Chemical Particles*"), if most subatomic particles in the universe were revealed, scientists would be shocked by how little they know and how little they have scratched the surface of the gigantic chemical iceberg.

The last generation of turbulent split-gathering was about the formation of the precursors of the smallest particles that will never be scientifically discovered. Indeed, as much as I believe that science has made tremendous efforts to help humankind understand nature, I also believe that some particles exist in nature and are smaller than the smallest particles currently known by science, and due to their size and the limits of scientific inquiries, some of them will never be detected by physical equipment. These particles may never be known by science. For instance, some spiritual substances belong to this class of particles. In the religious books I wrote on the creation of the universe, I explained how the precursors of angels can fit this generation of split-gathering. Sometimes, I find it shocking how a lot is believed about spiritual things (e.g., the prophetic, magic, miracles, witchcraft, occultism, and other forms of spiritual activities), yet some scientists refuse to address these issues in public, while some of them highly engage in spiritual practices secretly. Consequently, there are some truths that science will never unearth using traditional means, for they are hidden and unlockable or decryptable only by codes or languages that defy modern scientific methodology, which has neglected the spiritual aspects of the physical things it investigates.

Fig. 16: Hierarchy of the systems of bodies in the universe according to their inclusion levels

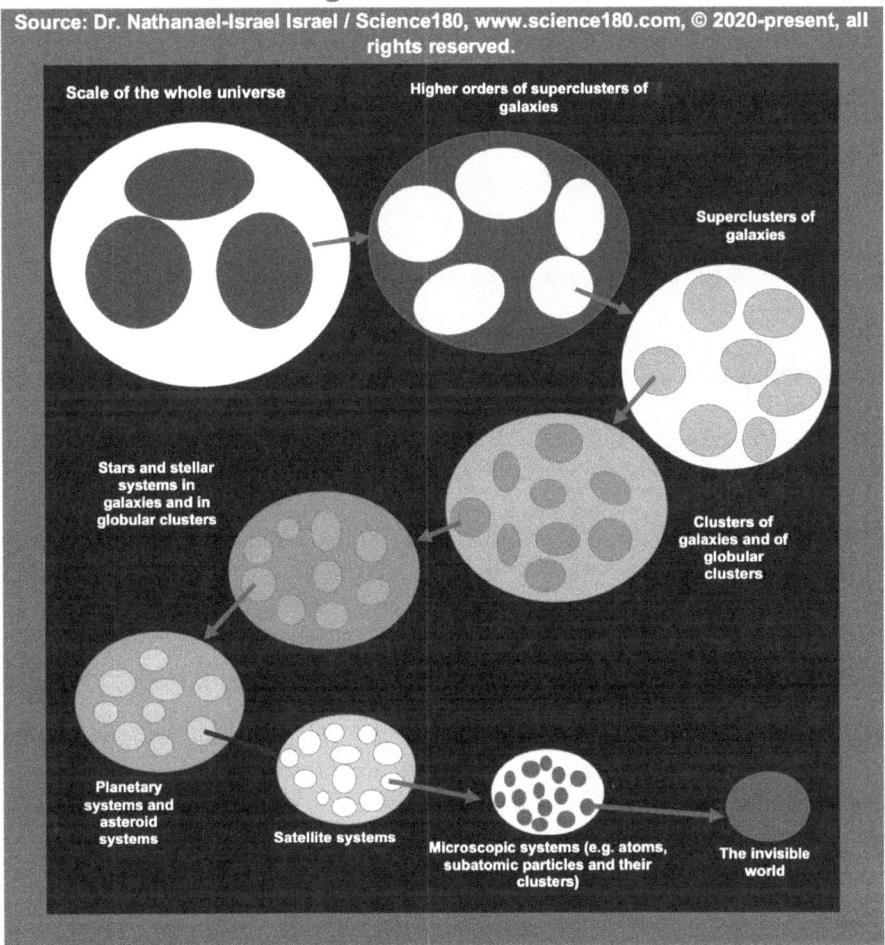

Because of the cascade of split-gathering of the precursor of the celestial bodies, the universe today appears as invisible things embedded in microscopic things (e.g., atoms), which are embedded in celestial bodies like satellites (orbiting planets), which are embedded in planetary systems (e.g., planets and satellites orbiting a star), which are embedded in stellar systems (e.g., stars orbiting a galaxy), which are embedded in galactic systems (e.g., galaxies orbiting a cluster of galaxies), which are embedded in galaxy clusters (e.g., clusters of galaxies orbiting a supercluster), which are embedded in higher orders of clusters and superclusters of galaxies, which, altogether, formed the whole universe as illustrated in Fig. 16, showing each level containing the next level.

Science180: Complex Universe-Origin, Life-Origin, and Chemicals-Origin Questions Solved Accurately in a Simple Language

FROM SCIENCE TO BIBLE'S CONCLUSIONS

At this point, I will address a phenomenon called intermittence. In simple terms, the intermittence of radius is the presence of smaller bodies, smaller clusters of matter, or smaller systems of bodies between bigger or larger ones. For instance, large planets are separated by small asteroids. Similarly, small clusters of stars and even isolated stars are usually found between galaxies and clusters of galaxies. The existence of small bodies between larger ones can be attributed to intermittence. The law that shaped the precursors of bodies in the universe did not always allow things to be split neatly, leaving some of their substance behind in the areas of their split. Those who do not understand this law always tend to explain the formation of larger bodies as the consequence of the collection of smaller ones or came to realize that, if there is anything that has caused the scientific community to fail to properly explain the origin of the universe and its content, it is also the ignorance or lack of understanding of the intermittence and its origin.

Applying the same mistake to living organisms, some people tried to explain the formation of larger organisms as the evolution of smaller ones, just as some people mistakenly try to explain the large celestial bodies as the evolution of smaller pieces of matter. Likewise, some people incorrectly try to explain the formation of all chemical elements as the product of chemical reactions between hydrogen atoms, which they erroneously think is the raw material from which all chemical elements were formed. The same error leads some people to believe that the formation of all animals follows a progression from one to the next. In this book, I reviewed all types of intermittences found in nature, but I focused on what the general public needs to know about the intermittence of celestial bodies. In my scientific book on the origin of the universe, I devoted many pages to explaining the concept of intermittence alone.

I have demonstrated that the size intermittence of celestial bodies in the universe can be explained by how the "cleavage" lines, "cleavage" points, interface of separation, or the interface of breakup of the precursors of the bodies were not neat or sharp, therefore leading to the formation of smaller bodies or small systems of bodies between larger ones. For instance, just as when you break a piece of cornbread in half, some pieces fall off, so, to some extent, when precursors of bodies were being broken during the formation of the universe, pieces of bodies came off, particularly at the breaking points or regions. Not only did small pieces come off, but they were also gathered into bodies, which ended up between larger ones. Sometimes, depending on the environment in which they formed and the characteristics of their precursors (e.g., viscosity, energy, and position), some large bodies formed without smaller ones between them, just as many clusters of small bodies formed without larger bodies near them. It all boils down to the law of the split-gathering of things. For instance, the Martian planetary system (the system formed by the planet Mars and its satellites) is very small and is found between the large Jovian planetary system (the system formed by the planet Jupiter and its satellites) and the relatively large Earth-Moon system (which is comprised of the Earth and its only satellite known as the Moon, which you see in the night sky). In each planetary system, smaller satellites are usually

Nathanael-Israel Israel: Acknowledged as Undisputable Specialist of all
Questions at the Intersection of Science and Biblical Creation

found between larger ones. As the precursors of the planetary systems were split-gathering, smaller amounts of fluids came off from the edges of their layers and then formed the precursors of other small bodies, including dust and microscopic particles found, for instance, in the interplanetary spaces, meaning the space between planets. The precursors of many asteroids were formed between the precursors of planetary systems, and they went through processes comparable to those of the precursors of the planets, except that their size was generally much smaller and their turbulence weaker. On scales smaller than that of celestial systems, intermittence is also observed. For instance, among the bigger rocks and minerals on Earth, smaller ones exist. Some complex rocks and minerals are separated by simpler ones. Likewise, although some minerals are pure and composed of the same chemical elements, particles collected or dug from mines, ores, and wells suggest that smaller clusters of minerals exist between bigger ones. Bigger chunks of atoms and molecules are mixed with smaller ones. Soil analyses show that many types of atoms and molecules are typically present. Hence, it is difficult to purify certain minerals and compounds and to remove impurities. Small atoms, molecules, and chemical compounds are also present between larger ones.

At this point, I would like to say a few words about the precursors of the atmosphere and space between celestial bodies. Indeed, celestial bodies are separated by a huge void. For instance, when we look at the night sky, we see stars separated by huge spaces. There is also a space between the planetary systems, asteroids, and satellites in the Solar System. Even on the atomic level, there is a space between the constituents of atoms. Why are there all of these spaces? To make a long story short, I would say that, when the fluid layers of the precursors of bodies were separating and moving from one location to another before splitting from neighboring precursors, a space appeared between them. To put it another way, after their separation, consecutive fluid layers became surrounded by space. During the formation of the celestial bodies, some of the space that formed between them became their atmospheres. In *"Turbulent Origin of Chemical Particles,"* I detailed how the atmospheres of the celestial bodies were formed. In the space separating celestial bodies, dust is found. And talking about dust in space, it may be worth mentioning that one of the major errors in some existing theories on the origin of the universe is the incorrect belief in the existence of a substance called "ether," allegedly thought to fill space, and which some people have used to try to explain gravity.

Finally, celestial bodies are not the only things in the universe. Chemical particles and living things are also present. Across the globe, most people also believe in and/or are aware of the existence of angels and demons. Even irreligious people believe in miracles and other mysterious demonstrations pertaining to the spiritual world. Therefore, I need to say a few words about the precursors of some spiritual things. As I demonstrated in other books (e.g., "Reconciling Science and Creation Accurately" and "Turbulent Origin of Life"), the turbulent prima materia (the original matter in the universe) did not go through many changes before its daughter bodies shaped the matter that was used to form spiritual beings and things. Because angels were made from a less-modified version of the turbulent prima materia, they move very fast (reaching billions of billions of miles in the blink of an eye). They can also transform themselves into other types of bodies, just as the turbulent prima materia could become anything. In other words, the ability of angels to change themselves into different shapes and to mimic other beings is related to the nature of the substance they were made of. Moreover, throughout my books, I declined to address the origin of God, for I personally believe that no one can explain where God came from or how He started to exist (if His existence had a beginning). Those are questions that those who will meet God someday may ask Him if they will still need to know them. In another book, I devoted tens of pages to the precursors of spiritual things.

4.4. Take-home message

Just as people want to know where they came from or what their roots are, they also want to know how the universe began, but they do not usually know how to find out. In this chapter, I showed that the bodies in the universe came from a bulk of substance that I called the turbulent prima materia, which appeared out of nothing and then was fragmented by a turbulent "explosion" (that I termed the original mysterious scattering), which broke it open, birthing various chunks of precursors, which were pushed into different directions, therefore unleashing the beginning of the expansion of the universe, which continues until today. I also explained that the turbulent prima materia was the substance used to fashion all types of matter or bodies in the universe.

As they were moving through space, some chunks, or pieces of the turbulent prima materia, also split-gathered into smaller bodies and so on and so forth, therefore contributing to the hierarchical organization of the universe into clusters of bodies on various scales. Variations in the conditions that existed during the bodies' formation by location contributed to the diversity of bodies in the universe. Because the original matter in the universe was rich in energy, the bodies formed out of it are also rich in energy. Hence, everything in the universe contains energy and particles capable of moving things. In other words, everything in the universe today is a modified version of the energetic particles originating from the turbulent prima materia. I also showed that, during the formation of the universe, turbulence occurred on different scales, with different intensities and at different positions, leading to the formation of various structures in the universe. Although a lot of data has been collected on countless particles and celestial bodies, a challenging task is explaining all the turbulence scales that exist in the universe and relating them to one another in a way that accounts for everything. This task, which I handled in my books, is like the genealogy, or ancestry line, of everything in the universe. This chapter elucidated how the precursors of bodies in the Solar System and other systems in the universe were formed. In the next chapter, I will detail the durations of some key processes in the formation of celestial bodies.

CHAPTER 5

USING A SIMPLE SCIENTIFIC FORMULA, CAN WE DETERMINE THE TIMELINE OF THE UNIVERSE'S FORMATION WITH ABSOLUTE CERTAINTY?

The age of the universe and the duration of the processes that formed it are hotly debated topics across the globe. Some people think that the process that shaped the universe took billions of years, while others believe it took just a few days. In this chapter, I will show you how, for the first time in history, I used scientific data to settle this disagreement. I will show you how I demonstrated the chronological order of key events during the universe's genesis. Because the formation of celestial bodies involved many processes, I had to break them down and scientifically assign a timescale to each based on the facts. Here, by timescale, I mean the time it took for a specific event to occur during the formation of the universe. By timeline, I mean the chronological arrangement of key events according to the order of their occurrence during the formation of the universe. In this chapter, for the sake of time and space, I will focus on the formation of the Earth, the Moon, the Sun, and other celestial bodies in the Solar System. After laying some fundamental principles concerning (1) the distance traveled by these celestial bodies before their fluid layers were gathered together as spherical bodies and (2) the speed of these travels, I will calculate the exact time it took for the universe to be formed. Please allow me to explain my thought process for addressing this challenging timeline question.

5.1. Principle for the calculation of the duration of the formation of celestial bodies

To demonstrate the duration of the processes that formed the universe, I relied on some of the 12 key facts I introduced earlier in this book, namely, the distances, speeds, and radii of the bodies. By the way, like I already said, the orbital speed, the radius, and the distance I used in this chapter and throughout this book were collected and approved by spatial agencies, including NASA, the US space agency. As I studied the universe, I understood that the speeds of celestial bodies and the distances separating them hold secrets about the duration of their formation. Indeed, as of today, if something or someone traveled a certain distance at a certain speed, the duration of that trip can be calculated by dividing the distance by the speed of the trip. I am pretty sure you have already done a similar calculation regarding the time it took you to travel a certain distance. For instance, if you drive 50 miles at 25 miles per hour, the time would be 2 hours (i.e., 50 miles divided by 25 miles/hour). I used the same logic to calculate the duration of the movement of celestial bodies from their initial position within their mother to their birth positions. For instance, by considering the distance separating key bodies and the speed with which their precursors moved, the duration of their movement can be estimated. I already showed that, as the precursors of the celestial bodies were forming, they also moved a certain distance before being gathered together.

Therefore, to estimate the time it could have taken for the universe to be formed, I focused on two main variables:

(1) the time it could have taken for the precursors of the bodies to be displaced from their initial position inside their mother's bodies and

(2) the time it could have taken for them to be molded into their current form once they reach their orbit.

Indeed, the precursors of the celestial bodies were not stationary or immobile when they were being shaped by the processes that split-gathered them. But they were moving and being reorganized. For instance, when we consider a house built by human beings, it takes time to gather the materials first and to put them together to build something. Depending on the location of the construction materials and how difficult it is to find and bring them to the construction site, it could take less time to build something than to gather the materials needed. Some construction companies travel many miles before finding a quarry where they can dig and extract gravel for the foundations of their buildings. But it can take a few hours to build some houses, while others can take days, weeks, months, or even years, depending on the home's style or design. In the same manner, it could have taken less time to fashion some celestial bodies than to displace their precursors beforehand. The formation of celestial bodies was complete when they were placed into their orbits. For one thing, it was for the bodies to be formed, but another thing was for them to be placed in orbit. As I mentioned earlier, celestial bodies are not separated by chance. For instance, I showed that in a planetary system, the precursor of the satellites escaped the precursor of the primary planet.

I also showed that, in the Solar System, the precursor of the bodies orbiting the Sun escaped the precursor of the Sun at about the Sun's escape velocity. After escaping the precursor of the Sun, the precursor of the bodies orbiting the Sun traveled a certain distance before forming the Earth, and that travel occurred at a certain speed. Likewise, after escaping the precursor of the Earth, the precursor of the Moon traveled a certain distance at a certain speed before being formed. Using the distances separating the Moon from the Earth and the Earth from the Sun, and the speed of their precursor's travel, I estimated the duration of their travel. I also explained that, at some point in the formation of the celestial bodies, their precursors were fluid layers. Because most major celestial bodies, as we know them today, have a spherical shape, it took a certain time for the fluid layers of their precursors to be gathered together into a sphere. I therefore needed to estimate the duration of that event as well. To summarize, the two key events I used to estimate the timeline of the formation of the celestial bodies are

(1) the duration of the travel of their precursor from their initial position in their mother precursor to their current position in orbit, and

(2) the duration of the gathering together of their fluid layers into a spherical celestial body.

The rest of this chapter is to walk you through this process for the Earth, the Moon, and the Sun, and then to show you how this process can be generalized for other celestial bodies in the universe.

Estimating the time it took for the precursors of bodies to leave their mother precursors (after splitting from them) and reach their orbits required an intimate understanding of the distance traveled and the speed of the bodies under the influence of the turbulence their precursors experienced. I previously explained that the semi-major axis of a celestial body is the average distance separating it from its primary body. I also showed that the precursor of the secondary bodies escaped from the precursor of their primary bodies at about the primary body's escape velocity. For instance, the precursor of the secondary bodies orbiting the Sun escaped the precursor of the Sun at about the escape velocity of the Sun. Therefore, the duration of the travel of the bodies from their position as they were leaving their mother precursor to their orbit can be estimated using the semi-major axis and escape velocity. For example, using the semi-major axis of the Earth (i.e., the average distance separating the Earth from the Sun) and the escape velocity of the Sun (look at velocity as a speed), I estimated the duration of the travel of the precursor of the Earth-Moon system from about the precursor of the Sun to about the position of the Earth. The same formula can be used for the Moon, using the Moon's semi-major axis and the Earth's escape velocity. NASA reported that the escape velocity of the Sun is 617.6 km/s. Therefore, based on my demonstration, the precursor of the bodies orbiting the Sun escaped the precursor of the Sun at about 617.6 km/s. This also suggests that the bodies orbiting the Sun could have been ejected from the precursor of the Sun at about 617.6 km/s. But let me be clear: by saying that the bodies orbiting the Sun escaped the precursor of the Sun, I am not saying that the bodies orbiting the Sun came out of the Sun.

Because the fluid layers of the precursors of celestial bodies were moved and reorganized under the influence of turbulence, turbulent structures developed in them. The most renowned structures formed in a turbulent fluid are vortices. These 3-D structures formed when fluid filaments or parcels swirled or wrapped around to gather together. In *"Turbulent Origin of the Universe*," I showed that a fluid layer moving at a certain speed under the influence of turbulence can swirl to collect itself into a 3-

D body just as a fluid filament can swirl to form a vortex. I showed that, just before the fluid layers of the precursors of the celestial bodies changed their shape from being almost linear to being spherical, their length was about the circumference of the celestial bodies born from them. In other words, as the fluid layers of the precursors of the celestial bodies were moving, they were collected until they became almost as long as the circumference of the bodies they would form, and then swirled to form those bodies. I coined the term "circumference timescale" to indicate the time it would have taken for the fluid layers of a body's precursor to swirl and form that body. I estimated that time by focusing on the body's circumference and orbital speed. It is the time that a particle (assumed to be spherical) takes to cover its circumference as it moves at its orbital speed.

As a reminder, the circumference of a circle is calculated by multiplying its diameter by 3.14 (also called pi and annotated π), meaning that the circumference of a body whose radius is R is $2\pi R = 2 \times 3.14 \times R$. By the way, in this formula, the x sign is the multiplication sign.

To ensure that everybody understands the meaning of the circumference timescale and how I calculated it, let me break it down a little bit. Imagine a cord of length "L" that you start rolling to form a circle. If you use the entire cord, the circumference is the distance around its circle, that is "L." Imagine you rolled that thread at an average speed V. The time it would take to roll that full circle would be the length divided by the speed (duration of time = L / V). In other words, knowing the circumference of a circle and the speed at which it was formed by rolling a cord or a linear rope, the time elapsed can be estimated. The same thing I just explained can be done for a sphere (think of it as a body with an orange-like shape). The circumference at the equator of such a body is like the maximum distance that can be traveled all the way around that body. In other words, if you take a sphere and use a cord in your hand to measure its circumference all the way around from the equator all the way to the pole, you will notice that the circumference decreases. A slice of orange in the middle is larger or longer than a slice toward its pole. To rephrase it, the circumference at the equator or at the middle of a sphere is like the maximum distance that can be traveled all the way around that sphere. For instance, if you remove the peel of an orange at its equator, open it, and flatten it into a cord or line, you will notice that the length of that slice can be roughly equal to the circumference. If you try to roll or curl that peel to make a circle, as I explained above, the time it will take to do it will be equal to the length of that slice (which was the circumference of the orange at its equator) divided by the speed of the curling motion. Trying to make myself very clear, the time it could take to move around the circumference of a sphere is equal to its circumference divided by the speed of the motion. Everything I explained in this segment so far also applies to celestial bodies: the time it would take to travel their circumference is equal to that circumference divided by the speed of travel.

In the case of the celestial bodies, their precursors were like fluid layers that ended up being spiraled or whirled around on their axes to form bodies, some of which are spherical. I extensively showed in my popular book *"Turbulent Origin of the Universe"*

68

that the fluid layers of the precursors were wrapped around, spiraled, or curled at about the orbital speed of the celestial bodies. Therefore, the time it would take for a near-linear fluid layer of length equal to the circumference of a body to roll at about the orbital speed can be estimated as the circumference divided by the orbital speed. Using the same mathematics that I explained above, the time it could have taken for the precursor of a celestial body to be rolled to form that body can be estimated as the ratio between its circumference and orbital speed. Because some precursors of bodies could have been rolled many times, the time to form them could be a little longer. Now, I will address the formation times of the Earth, the Moon, and the Sun.

5.2. Are we giving the formation of the Sun a bad interpretation scientifically or biblically?

Before the precursor of the Sun could be formed, the fluid of the precursor of the bodies orbiting the Sun had to escape first. I coined the term "escape time of the precursor of the Sun" to indicate the time needed for the fluid layers of the precursor of the bodies orbiting the Sun to escape the precursor of the Sun. It was after this escape that the precursor of the Sun was ready to swirl to form the Sun. For now, let us first estimate that time.

Because Mercury is the closest body to the Sun, I showed that the precursor of the bodies orbiting the Sun could not have traveled farther than the position of Mercury before it started to split. Dividing the semi-major axis of Mercury (57,910,000 km) by the escape velocity of the Sun (617.6 km/s), I showed that in 26.046 hours, the precursor of the bodies orbiting the Sun could have traveled from the position of the precursor of the Sun to the orbit of Mercury. If you would like to verify this math yourself, please remember to convert units properly when going from seconds to hours. For example, because there are 60 seconds in one minute and 60 minutes in one hour (meaning that one hour is 3600 seconds), to convert a duration in seconds into hours, you need to divide the duration in seconds by 3600.

Because the fluid layers of Mercury (the innermost bodies orbiting the Sun) would have been on top of the fluid layers of all the other bodies orbiting the Sun, about 26.046 hours after the beginning of the split of the precursor of the Solar System, the fluid layers of Mercury and the fluid layers of the other bodies orbiting the Sun must have reached the position of Mercury. This also means that about 26.046 hours after the beginning of the Solar System, the precursor of the Sun would have been properly formed and ready to start winding up to become the Sun. The duration of the formation of the Sun would have been the sum of the timescale I just calculated (26.046 hours) and the duration of time it took for the fluid layers of the precursor of the Sun to swirl and form the Sun once its precursor was formed. Let us now estimate that timescale.

My book *"Turbulent Origin of the Universe"* explained in more detail that by the time the precursor of the Sun had formed, the speed of its fluid layers was about the Sun's orbital speed (19.4 km/s). By the way, the orbital speed of 19.4 km/s is not measured by me, but by NASA and other space agencies. Considering the physics of how fluids can deform and form vortical bodies, I showed that the length of the fluid layers of the precursor of the Sun could have been about the circumference of the Sun at the moment these fluid layers swirled to form the Sun. Because the circumference of a body whose radius is R is $2\pi R = 2 \times 3.14 \times R$, I estimated that the circumference of the Sun is $2 \times 3.14 \times$ the Sun's radius. Because the Sun's radius is 696,000 km, the circumference of the Sun is $2 \times 3.14 \times 696{,}000$ km $= 4{,}370{,}880$ km. By dividing the circumference of the Sun by the orbital speed of the Sun (4,370,880 km / 19.4 km/s $= 62.58$ hours), I showed that about 62.58 hours after the precursor of the Sun was formed, it wrapped around to form the Sun. Because it took about 26.046 hours before the precursor of the Sun was formed, considering what I said in the previous sentence, the Sun was formed 88.63 hours (i.e., 26.046 hours + 62.58 hours) after the beginning of the split-gathering of the precursor of the Solar System. Because one day is 24 hours, 88.63 hours is 3.693 days, meaning that the Sun was formed on the 4th day after the beginning of the formation of the Solar System. In summary, the formula I used to estimate the date of the formation of the Sun with respect to the beginning of the formation of the Solar System is (Fig. 17):

Birthdate of the Sun = "Semi major axis of Mercury" / "Escape velocity of the Sun" + 2 x 3.14 x (Radius of the Sun) / "Orbital speed of the Sun"

Nathanael-Israel Israel: Known as the #1 Universe-Origin, Life-Origin, and Chemicals-Origin Scientist & Mathematician

Fig. 17: Formula of the Birthdate of the Sun

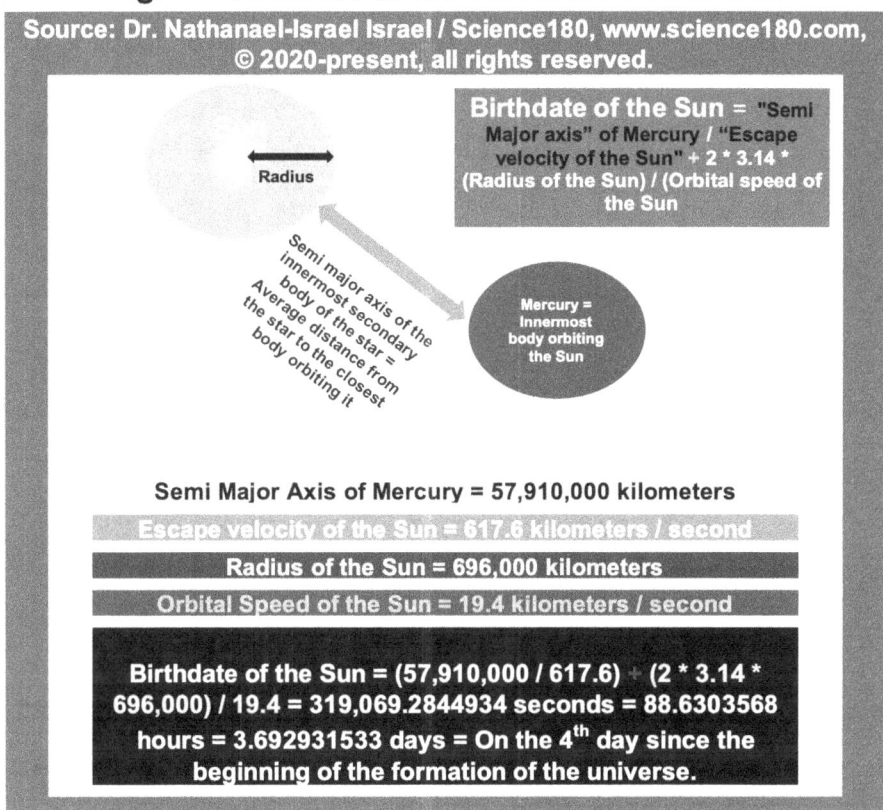

Semi Major Axis of Mercury = 57,910,000 kilometers

Escape velocity of the Sun = 617.6 kilometers / second

Radius of the Sun = 696,000 kilometers

Orbital Speed of the Sun = 19.4 kilometers / second

Birthdate of the Sun = (57,910,000 / 617.6) + (2 * 3.14 * 696,000) / 19.4 = 319,069.2844934 seconds = 88.6303568 hours = 3.692931533 days = On the 4th day since the beginning of the formation of the universe.

'Science180 Academy' Success Strategy
SCIENCE180 SEMINARS

People whose awareness is raised by Science180 usually ask me to go deeper or they wonder "what's else?". That is one of the reasons Science180 trains them through strategic work sessions (during seminars or training sessions) that transfer customizable skills and solutions to them. Science180 Seminars are client-centered and tailored to strongly engage the clients so they maximize the discovery of and the tapping into new opportunities, and exponentially outperform their expectations. Science180 offers customizable seminars that can be labeled as a colloquy, conference, consultation, discussion, forum, keynote speech, lecture, lesson, meeting, symposium, summit, study group, tutorial, workshop or working section accordingly on any topic related to:

- Universe-origin for scientists and mathematicians, philosophers, laypeople, and the general public
- Universe-origin or universe creation for believers
- Life-origin for life scientists, for all other scientists, and for believers
- Chemical-origin for scientists
- Universe-origin seminars for children
- Universe and life-origin for pseudepigraphic believers

As you contact us with your needs, we can customize your program accordingly. Learn more at Science180Seminars.com.

5.3. How can people scientifically approach difficult topics such as challenging the accuracy of the Biblical creation, the Big Bang, or any theory on the duration of the formation of the Earth?

I showed that in the beginning of the Solar System, the precursor of the bodies orbiting the Sun escaped the precursor of the Sun, and then traveled for about 149,600,000 km (i.e., the semi-major of the Earth, which is the average distance separating the Sun and the Earth) before the precursor of the Earth-Moon system split from the stack of fluid layers carrying all the precursors of the bodies orbiting the Sun. Because of the distance traveled by the precursor of the Earth-Moon system inside the fluid layers of the precursor of the bodies orbiting the Sun, the duration of time that elapsed since the beginning before the precursor of the Earth-Moon system was formed is about the distance separating the Sun and the Earth divided by the escape velocity of the Sun. I hope you understand this formula; if not, let me explain again. Indeed, because the precursor of the Earth-Moon system traveled for about the distance separating the Sun and Earth at about the escape velocity of the Sun, the duration of that travel equates to the distance divided by the speed of that travel. Hence, that timescale is:

CHAPTER 5: TIMELINE OF THE FORMATION OF THE UNIVERSE

$$149{,}600{,}000 \text{ km} / 617.6 \text{ km/s} = 67.286 \text{ hours} = 2.804 \text{ days}$$

This duration of time is what I call the "semi-major axis timescale" of the Earth.

After the precursor of the Earth-Moon system split from the stack of fluid layers of the precursor bodies orbiting the Sun, it quickly split into the precursors of the Earth and the Moon. The precursor of the Earth is like the leftover of the precursor of the Earth-Moon system after the precursor of the Moon escaped (or was stripped away from) the precursor of the Earth. It was like, when the precursor of the Earth-Moon system was about to swirl, the precursor of the Moon was stripped away, and the leftover, the precursor of the Earth, continued its motion and formed the Earth. I lengthily demonstrated that within a few minutes (precisely in less than 30 minutes), the precursor of the Moon escaped the precursor of the Earth, therefore paving the way for each of these precursors to collect their fluid layers into individual bodies (for the details, please refer to my book *"Turbulent Origin of the Universe"*). In other words, the precursor of the Earth and the precursor of the Moon were formed as individual bodies about 67.286 hours, or 2.804 days, after the beginning of the formation of the Solar System.

Just as I showed for the precursor of the Sun, the fluid layers of the precursor of the Earth moved and, upon reaching a length of about the Earth's circumference, they swirled to form the Earth. Knowing the radius of the Earth, I calculated the circumference of the Earth as:

$$2 \times 3.14 \times 6378.137 \text{ km} = 40054.7 \text{ km}$$

The fluid layers of the precursor of the Earth swirled at about the orbital speed of the Earth (29.78 km/s) and formed the Earth. Dividing the circumference of the Earth by the orbital speed of the Earth, I found that the swirling lasted about 22.42 minutes (i.e., 40054.7 km / 29.78 km/s):

$$40054.7 \text{ km} / 29.78 \text{ km/s} = \text{about } 22.42 \text{ minutes}$$

These 22.42 minutes are what I called the "circumference timescale" of the Earth. Because the precursor of the Earth was formed about 67.286 hours after the beginning of the Solar System (i.e., semi-major axis timescale of the Earth) and because it took about 22.42 minutes for the fluid layers of the precursor of the Earth to swirl and form the Earth (i.e., circumference timescale of the Earth), the total amount of time that elapsed from the beginning of the formation of the Solar System until the Earth was formed was:

$$\text{Birthdate of the Earth} = 67.286 \text{ hours} + 22.42 \text{ minutes} = 67.66 \text{ hours}$$
$$= 2.82 \text{ days}$$

To put it another way, the formation of the Earth was finalized on the 3rd day after the beginning of the formation of the Solar System. The formula for the date of the formation of the Earth can be summarized as (Fig. 18):

Birthdate of the Earth = "Semi major axis of the Earth" / "Escape velocity of the Sun" + 2 x 3.14 x (Radius of the Earth) / "Orbital speed of the Earth"

Fig. 18: Formula of the Birthdate of the Earth

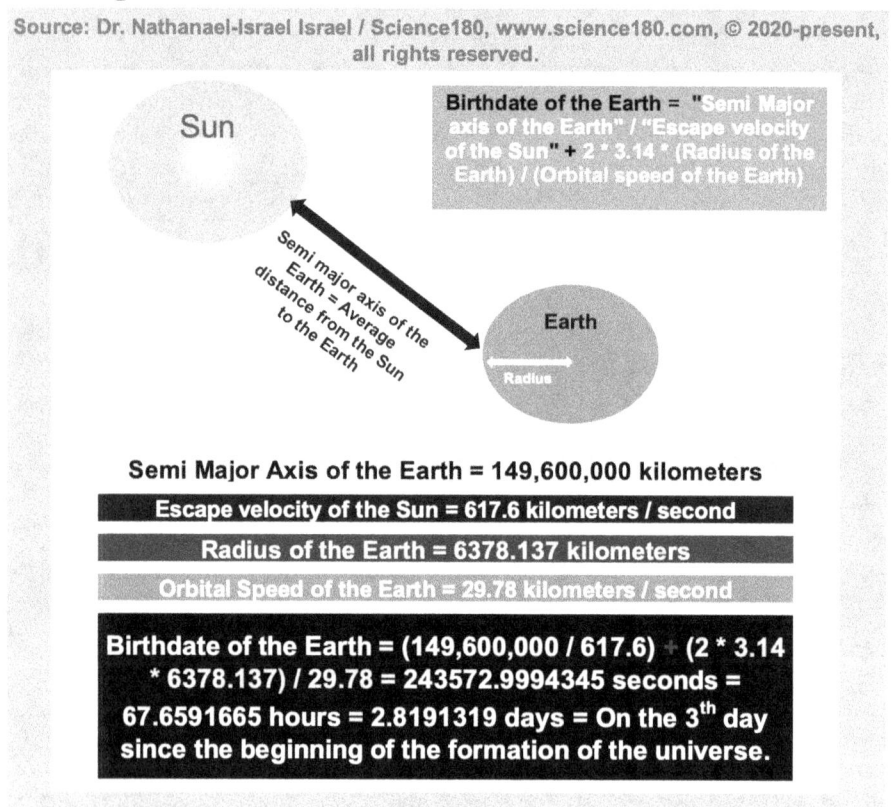

Nathanael-Israel Israel: Known as the #1 Universe-Origin, Life-Origin, and Chemicals-Origin Scientist & Mathematician

'Science180 Academy' Success Strategy
SCIENCE180 CONSULTING

Because Science180's trainings, seminars, or strategic work sessions (through which it transfers skills and training solutions) are great, some customers want to go even deeper on a long-term, sustainable basis. That is where Science180 Consulting, one-on-one consulting, and mentoring (that some people may prefer calling coaching programs) comes in. That is where Science180 can truly change people's behavior on a long-term basis according to their specific needs. With Science180 Consulting, you will discover and understand the deep secrets of the formation of the universe, life, and chemicals around you. Hear Dr. Nathanael-Israel Israel's personal selection and teaching on key topics that will help you break the code of the universe formation and functioning. All strategically designed to enlighten you, guide you to navigate and filter the massive data collected on the universe and its content so you know how to answer the world's most challenging origin questions, remove any scientific and philosophical cataracts that may be blocking you, and help bring you many steps closer to your best life today and forever. Science180 Consulting will train you, transfer unconventional skills to you and change your behavior so you go deeper. To get started today or to learn more, go to Science180Consulting.com.

5.4. Can God scientifically help you to rationally debunk the duration of the Moon formation taught at top Western universities? This scientist says YES!

Like I already demonstrated, the precursor of the Moon descended from the precursor of the Earth-Moon system. In fact, as the precursor of the Earth-Moon was reorganizing itself, the precursor of the Moon escaped the precursor of the Earth. Another way to say this is that the Moon's precursor was ejected from Earth's precursor. In *"Turbulent Origin of the Universe,"* I showed that the split of the precursor of the Earth-Moon system into the precursor of the Earth and the precursor of the Moon was very fast and lasted a few minutes. I also showed that the precursor of the Moon escaped the precursor of the Earth at about 11.186 km/s (i.e., the Earth's escape velocity). Then, it traveled about 384,400 km (i.e., the semi-major axis of the Moon = the distance separating the Earth and the Moon) before being ready to swirl to form the Moon. By dividing the distance between separating the Earth and the Moon by the escape velocity of the Earth (i.e., the speed with which the precursor of the Moon escaped the precursor of the Earth), I demonstrated that 9.54 hours (384,400 km / 11.186 km/s) after the precursor of the Moon escaped the precursor of the Earth, it was set to swirl and form the Moon.

$$384{,}400 \text{ km} / 11.186 \text{ km/s} = 9.54 \text{ hours}$$

I will now calculate how long it took for this swirling to occur.

Indeed, due to the turbulence that was occurring inside the fluids of the precursor of the Moon, the length of this precursor was reduced to about the circumference of the Moon just before this precursor swirled to form the Moon. Knowing the radius of the Moon (1738.1 km), the circumference of the Moon is:

$$2 \times 3.14 \times 1738.1 \text{ km} = 10915.27 \text{ km}$$

Based on my previous demonstrations, the precursor of the Moon swirled at the Moon's orbital speed (1.02316 km/s). To calculate the duration of that swirling, I divided the circumference of the Moon by the orbital speed of the Moon:

$$10915.27 \text{ km} / 1.02316 \text{ km/s} = 2.96 \text{ hours}$$

Because before taking 2.96 hours to swirl around, the precursor of the Moon took 9.54 hours to travel the distance separating the Earth and the Moon, I estimated that the total amount of time that elapsed from the moment the precursor of the Moon was formed until the Moon was formed was:

$$9.54 \text{ hours} + 2.96 \text{ hours} = 12.5 \text{ hours}$$

In other words, after the precursor of the Earth-Moon system split into the precursor of the Earth and the precursor of the Moon, about 12.5 hours elapsed before the Moon was formed. In the section on the formation of the Earth, I showed that the precursor to the Earth-Moon system formed about 67.286 hours after the beginning of the Solar System's formation. Because the Moon was formed about 12.5 hours after the formation of the Earth-Moon system, I deduced that the Moon was formed 79.786 hours (i.e., 67.286 hours + 12.5 hours) after the beginning of the formation of the Solar System. Because one day is 24 hours, 79.786 hours equals 3.324 days (i.e., 79.786 hours / 24 hours/day).

Birthdate of the Moon = 67.286 hours + 12.5 hours = 79.786 hours = 3.324 days

In other words, the Moon was formed on the 4th day after the beginning of the formation of the Solar System. In summary, the formula to calculate the date of birth of the Moon with respect to the beginning of the formation of the Solar System is (Fig. 19):

Birthdate of the Moon = "Semi major axis of the Earth" / "Escape velocity of the Sun" + "Semi major axis of the Moon" / "Escape

velocity of the Earth" + 2 x 3.14 x (Radius of the Moon) / "Orbital speed of the Moon"

Fig. 19: Formula of the Birthdate of the Moon (Earth's Satellite)

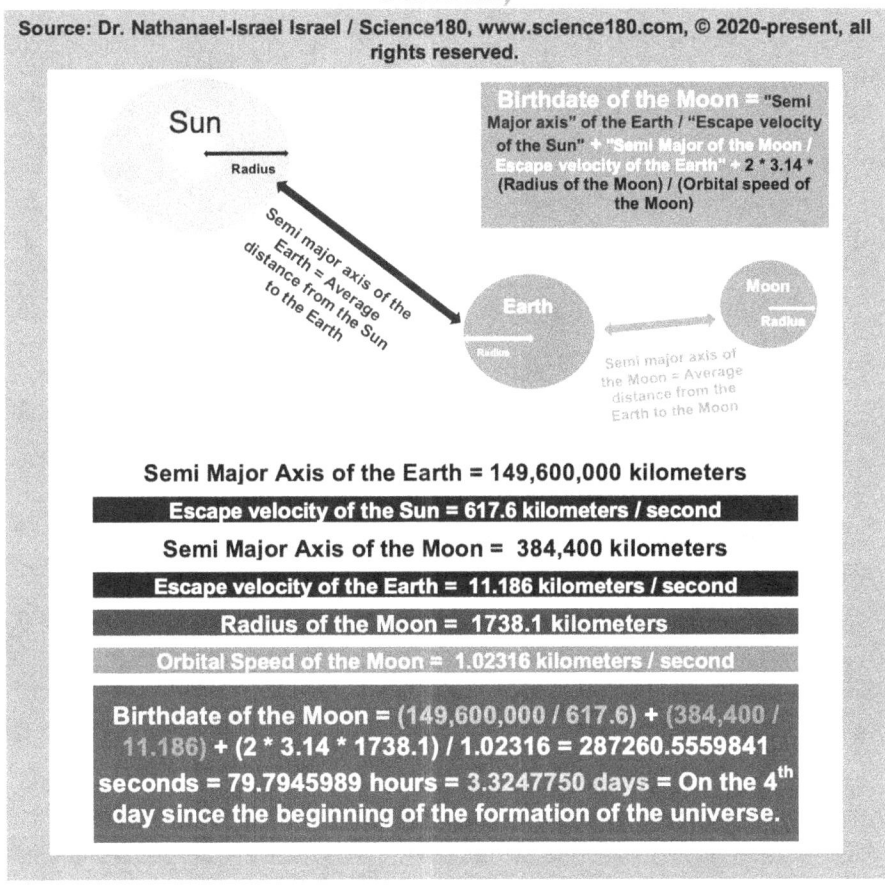

Semi Major Axis of the Earth = 149,600,000 kilometers

Escape velocity of the Sun = 617.6 kilometers / second

Semi Major Axis of the Moon = 384,400 kilometers

Escape velocity of the Earth = 11.186 kilometers / second

Radius of the Moon = 1738.1 kilometers

Orbital Speed of the Moon = 1.02316 kilometers / second

Birthdate of the Moon = (149,600,000 / 617.6) + (384,400 / 11.186) + (2 * 3.14 * 1738.1) / 1.02316 = 287260.5559841 seconds = 79.7945989 hours = 3.3247750 days = On the 4th day since the beginning of the formation of the universe.

Another Book by Nathanael-Israel Israel:
SCIENCE180 ACCURATE SCIENTIFIC PROOF OF GOD

THE FIRST AND THE ONLY SCIENTIFIC BOOK THAT TALKS TO ANTI-CREATIONISTS, EVOLUTIONISTS, BIG BANG PROPONENTS, ATHEISTS, AND ALL OTHER FREETHINKERS AND RATIONALISTS ABOUT THE UNIVERSE FORMATION AND THEY BEG TO KNOW MORE ABOUT GOD, THE CREATOR, THAT THEY DENY.

As you read this historic book, you will:

- Scientifically know what is the one clear sign you should always pay attention to in your efforts to decipher the primary cause and the key drivers of the fundamental processes responsible for the universe-formation
- Discover the only way to scientifically know if God exist, and if so, which of the thousands of beings worshipped across the globe is the true God
- Accurately answer the most critical universe-origin and life-origin questions so you can stop standing in tension with consequential question marks including those related to religion and reason or the so-called war between science and the Bible
- Discover the errors in the scientific and religious theories about the universe-origin and life-origin that are putting you at a high risk you will never recover from if you don't quickly and confidently learn how to rationally take control over threats lurking at the edge of your efforts to understand the universe and life today
- Challenge the cosmological status quo and embrace the real change that will disrupt the hidden cages that may be holding you and that you ignore
- Definitively answer all your doubts about the source or author of the universe and life … (learn more at Science180.com/godproof)
- Understand that religion or faith, reason or science can coexist and can be properly reconciled to accurately lead you to the correct source of everything in the universe
- Satisfy your burning desire for freedom from beliefs and scientific theories about the universe-origin and life-origin that suffocate you and bind your mind, faith, unbelief, heart, and education
- Scientifically set on fire all false theories or dogmas about the existence of God, the Creator, that are enslaving humankind

Whether you are a believer, unbeliever, freethinker, administrator, politician, curriculum designer, curriculum specialist, education policymaker, teacher, librarian, school board member, researcher, parent, student, clergy, or a layperson, as long as you are really seeking to scientifically understand the

Nathanael-Israel Israel: Known as the #1 Universe-Origin, Life-Origin, and Chemicals-Origin Scientist & Mathematician

rational proof of the existence of God, *"Science180 Accurate Scientific Proof of God"* is the much-admired book written for great people just like you! Grab your copy today and start reading it! Don't wait any longer!

Dr. Nathanael-Israel Israel is a Beninese-American scientist, entrepreneur, and international consultant, who shows people of all ages and educational backgrounds how to scientifically decode the formation of the universe and of life, and who is acknowledged as the creator of the Chemicals Turbulent Origin Formula™, the inventor of the Life Turbulent Origin Formula™, and the discoverer of the Universe Creation Formula™. He is the Founder of Science180 Academy, which is trailblazing the reconciliation between science and the creation.

5.5. Why you don't have to scientifically check out your brain to rationally prove the intersection of the Biblical creation and the scientific evidence

Here, you will scientifically see whether or not the Biblical account of creation matches the scientific evidence. Indeed, the calculation I did above concerning the formation dates of the Earth, the Moon, and the Sun provided scientific evidence for these dates and ways to assess creation stories. Most of the religions in the world have stories of the formation of the universe. In my popular book called *"Science180 Accurate Scientific Proof of God,"* I detailed the creation stories of the following religions or ideologies: animism, Buddhism, Confucianism, Hinduism, Islam, and evolutionism. After reviewing these creation narratives, I realized that none of them specifically detailed the timeline of the formation of the Earth, the Moon, and the Sun as the Bible did. Therefore, for the sake of space, I will just compare the Biblical account of creation with the scientific evidence. Because, as they read this book, some people who are not familiar with the Biblical story of creation may not have a Bible by them to check out what the Bible's Book of Genesis (written about 3,500 years ago) said about the beginning of the Universe, I will quote it here before I talk about it:

The Bible's Book of Genesis 1:1: "In the beginning, God created the heaven and the earth. 2 And the earth was without form, and void; and darkness was upon the face of the deep. And the Spirit of God moved upon the face of the waters. 3 And God said, Let there be light: and there was light. 4 And God saw the light, that it was good: and God divided the light from the darkness. 5 And God called the light Day, and the darkness he called Night. And the evening and the morning were the first day. 6 And God said, Let there be a firmament in the midst of the waters, and let it divide the waters from the waters. 7 And God made the firmament and divided the waters which were under the firmament from the waters which were above the firmament: and it was so. 8, and God called the firmament Heaven. And the evening and the morning were the second day. 9 And God said, Let the waters under the heaven be gathered together unto one place, and let the dry land appear: and it was so. 10 And

God called the dry land Earth; and the gathering together of the waters called the Seas: and God saw that it was good. 11 …12 …13… And the evening and the morning were the third day. 14 And God said, Let there be lights in the firmament of the heaven to divide the day from the night; and let them be for signs, and for seasons, and for days, and years: 15 And let them be for lights in the firmament of the heaven to give light upon the earth: and it was so. 16 And God made two great lights; the greater light [Sun] to rule the day, and the lesser light [Moon] to rule the night: he made the stars also. 17 And God set them in the firmament of the heaven to give light upon the earth, 18 And to rule over the day and over the night, and to divide the light from the darkness: and God saw that it was good. 19 And the evening and the morning were the fourth day" (King James Version).

Because in the Biblical story of creation, the Bible clearly addresses the date of the formation of the Earth, the Moon, and the Sun, I will focus my review of the biblical narrative on these three celestial bodies before expounding my thoughts on other celestial bodies in the universe. Indeed, the Bible mentioned that waters started to be divided from waters on the second day, meaning after the first 24 hours. I showed scientifically that, 26.05 hours after the beginning, the precursor of the bodies orbiting the Sun split from the precursor of the Sun and began splitting into precursors of celestial bodies. This means that the scientific evidence supports the view that, by the second day after the beginning of the Solar System, fluid layers began to separate from one another. In other words, as the Bible says about the second day, the separation of fluids really began then. Scientific evidence also showed that, after the precursor of the bodies orbiting the Sun escaped the precursor of the Sun at about 617.6 km/s (i.e., escape velocity of the Sun), it was organized as a stack of fluid layers, including the fluid layers of the precursor of the Earth-Moon, which had to "wait" until all the fluid layers above it (e.g., the precursors of Mercury, Venus, and many asteroids located between the Sun and the Earth) split and were removed before it could (in its turn) split and gather itself into the precursor of the Earth and the precursor of the Moon.

I showed that the distance traveled by the precursor of the Earth-Moon system before it split from the fluid layers below it was about the semi-major axis of the Earth (i.e., the average distance separating the Sun and the Earth): 149,600,000 km. I also showed that the speed at which the precursor of the Earth-Moon system traveled from the precursor of the Sun to the location of its split from the stack of fluid was about the escape velocity of the Sun, 617.6 km/s. Dividing the aforementioned distance by that speed (i.e., 149,600,000 km / 617.6 km/s = 67.29 hours), I proved that the time it took for the precursor of the Earth-Moon system to travel that distance is about 67.29 hours, meaning 2.804 days:

$$149{,}600{,}000 \text{ km} \ / \ 617.6 \text{ km/s} = 67.29 \text{ hours} = 2.804 \text{ days}$$

I showed that the precursor of the Earth-Moon system split very quickly, in a matter of a few minutes, and then swirled at about the Earth's orbital speed (29.78 km/s) to form a spherical planet called Earth, whose equatorial radius is 6378.137 km. By dividing the circumference of the Earth by the orbital speed of the Earth, I demonstrated that 22.42 minutes after they reached the position of the Earth's orbit, all the fluid layers of the precursor of the Earth were gathered together to form the Earth. By adding (1) the 67.29 hours it took for the fluid layers of the precursor of the Earth-Moon system to move from about the precursor of the Sun to the orbit of the Earth, and (2) the 22.42 minutes it took for it to be gathered together around its circumference, I demonstrated that the duration of the formation of the Earth is:

Earth's birthdate = 67.29 hours + 22.42 minutes = 67.66 hours = 2.82 days

Hence, the formation of the Earth was completed on the 3rd day of creation, as recorded in the Bible's Book of Genesis (Genesis 1:9-13).

I also proved that the precursor of the Moon escaped the precursor of the Earth at about the escape velocity of the Earth (11.186 km/s) and traveled about 384,400 km (the distance separating the Moon and the Earth, also called the semi-major axis of the Moon) before reaching a point where it was collected into the Moon. By dividing the distance separating the Earth and the Moon (384,400 km) by the speed of travel (11.186 km/s), I showed that, after traveling for 9.54 hours away from the precursor of the Earth, the precursor of the Moon was ready to collect itself into a satellite.

384,400 km / 11.186 km/s = 9.54 hours

I proved that the time it took for the fluids of the precursor of the Moon to gather themselves into the Moon (radius = 1738.7 km) once they reached the orbit of the Moon was about the circumference of the Moon divided by the orbital speed of the Moon:

2 x 3.14 x 1738.1 km / 1.02316 km/s = 2.96 hours

Therefore, the time elapsed before the Moon was fully formed is the sum of (1) the 9.54 hours it took for the precursor of the Moon to escape the precursor of the Earth-Moon system and reach the orbit of the Moon and (2) the 2.96 hours it took for the fluid layers of the precursor of the Moon to be wrapped around to form the Moon after reaching its orbit:

9.54 hours + 2.96 hours = 12.5 hours

FROM SCIENCE TO BIBLE'S CONCLUSIONS

Because it took 67.286 hours for the precursor of the Earth-Moon system to move from the precursor of the Sun to a position in space where it split from the stack of fluids of the bodies orbiting the Sun and then split into its daughter bodies (the precursor of the Earth and that of the Moon), the duration of the formation of the Moon with respect to the Sun must take into account all of the time that occurred before the precursor of the Moon was even born:

Moon's birthdate = 67.286 hours + 12.5 hours = 79.786 hours = 3.324 days

This confirms that the Moon was formed on the 4th day, as the Bible said more than 3,500 years ago (Genesis 1:14-19).

Now, let's recap the timeline of the formation of the Sun. Indeed, I also elucidated that, before the precursor of the Sun could fully be formed and set to start gathering its fluid into a spherical body, the precursor of the bodies orbiting the Sun could have traveled a distance no higher than the average distance between the Sun and Mercury (Mercury being the closest body to the Sun), a distance called the semi-major axis of Mercury: 57,910,000 km. Because the precursor of the bodies orbiting the Sun escaped the precursor of the Sun at about 617.6 km/s (escape velocity of the Sun), I showed that the time it could have taken for the fluid layers of all the bodies orbiting the Sun to escape the precursor of the Sun and reach the position of Mercury was about:

57,910,000 km / 617.6 km/s = 26.046 hours

Once the precursor of the bodies orbiting the Sun escaped, the fluid layers of the precursor of the Sun moved at a speed near the orbital speed of the Sun (19.4 km/s) and swirled to gather together into a spherical and massive star called the Sun, whose radius is 696,000 km. Knowing the radius of the Sun, I estimated its circumference (2 x 3.14 x 696,000 km = 4,370,880 km), which is about the distance a fluid parcel could have traveled to complete one turn around the Sun. Dividing the circumference of the Sun (4,370,880 km) by its orbital speed (19.4 km/s) allowed me to calculate the time it took for the precursor of the Sun to gather together all its fluids to form the Sun once the fluid layers of the bodies orbiting it escaped:

2 x 3.14 x 696,000 km / 19.4 km/s = 4,370,880 km / 19.4 km/s = 62.58 hours

Therefore, the amount of time it took for the Sun to form is the sum of the 26.046 hours required for all the bodies orbiting it to clear the way or to escape the precursor of the Sun and the 62.58 hours needed for the fluid layers of the precursor of the Sun to swirl and form the Sun:

26.046 hours + 62.58 hours = 88.63 hours = 3.693 days

In other words, the Sun was formed on the 4th day after the beginning of the universe. This scientific evidence aligns perfectly with the Biblical creation narrative, which states that the Sun was formed on the 4th day (Genesis 1:14-19).

Considering the timeline I calculated for the formation of the Earth, the Moon, and the Sun, for the first time in history, I (Nathanael-Israel Israel) scientifically demonstrated that, just as the Bible said and many Jews and Christians have believed throughout the ages, the Earth was formed on the 3rd day, and the Moon and the Sun on the 4th day after the beginning of the universe. Although the origin of the universe has preoccupied human beings for a long time, and that, from science to philosophy, passing by mathematics (which some people do not consider as science because of the inability to experimentally verify some mathematical conclusions), people have tried to address the universe's origin scientifically and philosophically, and that many other books in the Bible talked about creation (e.g., the Book of Job, the Gospel of John, the Psalms of King David, the Epistle of Paul to the Ephesians, the Epistle of Paul to the Colossians, the Epistle of Peter, the Book of Revelation), none of these books is as detailed as the Genesis book, and no human being has ever scientifically demonstrated the timeline of the formation of the universe as I did. I considered it a privilege to be given the grace to do this demonstration, and I am grateful for what I shared with the world. Many other details surround the timeline story I presented in this chapter, but for the sake of space, I would like to refer you to my books called *"Turbulent Origin of the Universe"* and *"Reconciling Science and Creation Accurately*," where I elaborated on mysterious secrets hidden in the creation narrative, but that people have failed to pay attention to. For instance, the movement of the Spirit of God over the waters and the separation of waters from waters mentioned in the Biblical account of creation in the Book of Genesis allude to turbulence and the separation of fluids. Moreover, looking at the cascade of scattering of fire by some fireworks, I felt like the "explosive" cascades of breakups that birthed the precursors of stars could have been an astronomical display of gigantic fireworks. I also showed that the origin, or temporal reference, of the Solar System's formation was very close to the beginning of the universe. In other words, although all the stars in the universe could not have been born at the same time, due to the cascade of "explosive" events that birthed them, the origin or formation of the stars could have been fairly near the origin of the formation of the Solar System. Stars are separated by so much distance because of the speed of the events that birthed them and because of the gigantic size of the precursor of the universe.

Before I close this segment, it is important that I elaborate a little bit on the significance of the demonstration I did in this chapter. Unlike most of the scientific studies on the origin of the universe, which seem to focus on mathematical modeling and philosophical concepts that cannot be experimentally verified in a laboratory setting, while a vast amount of already-collected data is not properly accounted for to explain the unanswered questions, the Bible revealed things thousands of years ago that are now proven to be true. Above all, what is more striking and consequential is that, in the Biblical account of creation, Moses did not say that the universe was formed by chance but by God. Moses, the author of the Biblical creation narrative, was born around 1500 BC. Because the story Moses told millennia ago has been scientifically confirmed today, we must also accept that Moses said God is the Creator. Furthermore, in addition to the Book of Genesis, Moses authored 4 other books: Exodus, Leviticus, Numbers, and Deuteronomy. In those books, Moses repeatedly revealed that there is only one God, the God of Israel, referred to in the Biblical creation story as Elohim, also known to the Jews by many other names: Adonai, El, Yah, Yahweh, Jehovah, etc. In other words, although people across the globe and throughout history tend to label their idols as "gods," there is only one God, the God of Israel, who created the heavens and earth, and everything else in the universe. Any other being called a "god" or even "God" besides Him is a false god or an idol. It is a huge mistake to think or pretend to unite people across the globe by claiming that there are many gods or that the being called "god" in all the religions of the world is the same. In other words, although there are many "gods" or "idols," there is ONLY ONE GOD, the God Moses talked about and who is also known as the God of Abraham, Isaac, and Jacob. I would like for you to seriously take what I said in this segment about creation and reconsider your ways concerning who you believe is the creator of the world and how to obey and serve Him, for the day is coming when everybody will give an account of their own life and belief. Because the issues I addressed in my books on the origin of the universe go beyond science, this chapter would have missed the mark if I had not addressed their significance with respect to eternity.

'Science180 Academy' Success Strategy:
SCIENCE180 ACADEMY PROGRAMS

Owned by Science180, Science180 Academy is a training, speaking, consulting, and mentoring program specialized in everything universe-origin, life-origin, chemicals-origin, and anything at the intersection of reason and faith, or science and religion.

Science180 Academy deals with different subjects according to the needs of its members or target groups. When people register to Science180 Academy, they must choose the program(s) they want to focus on so their training can be properly personalized accordingly. This is similar to how people register to a university, and take classes in a specific department matching their needs!

Science180's breakthroughs are so complex and dense that it is not realistic or good to try to explain all in just one academy, else people will be overwhelmed, disinterested, and confused by the plethora of data to handle. In other words, Science180 Academy offers a wide range of origin-related training in various domains strategically designed to allow people to choose the most suitable for their needs so that, regardless of their background or field of

84

Nathanael-Israel Israel: Known as the #1 Universe-Origin, Life-Origin, and Chemicals-Origin Scientist & Mathematician

expertise, people can equip themselves, align their mindset, improve lives today and forever using the accurate explanation of the origin of the universe, of life, and of chemicals. Science180 Academy curriculum is based on 12 years of deep unconventional research that culminated with the publication of many much-admired books on the formation of the universe and its content (see www.Science10.com/books).

The content of each Science180 Academy is strategically crafted by Dr. Nathanael-Israel Israel (who is acknowledged as the internationally-acclaimed world's authority in origin-related issues) to suit both scientists and nonscientists, religious and nonreligious people, leaders as well as followers, so they can fully decode the proofs of the formation of the universe, of life, and of chemicals they have been wanting to demonstrate or grasp. The current programs of Science180 Academy (www.Science180Academy.com) are:

1. SCIENCE180 ACADEMY OF COSMOLOGY (Designed for all scientists who want to scientifically study cosmology, the science of the origin and fate of the universe)

2. SCIENCE180 ACADEMY OF TURBULENCE (This is a perfect fit for scientists and other experts interested in studying abiotic turbulence). Examples of these people include:

3. SCIENCE180 ACADEMY OF LIFE SCIENCES (Tailored to those who want to study biotic turbulence):

4. SCIENCE180 ACADEMY OF CHEMISTRY (Designed for chemists, biochemists, scientists, and other educated people who want to understand the origin of chemical particles)

5. SCIENCE180 ACADEMY FOR LAYPEOPLE OR THE GENERAL PUBLIC (Very fit for any layperson or "less" educated people who wants to learn (in a simple language) deep insights that even those who went to university for years were unable to decrypt by themselves, so these laypeople can be equipped to eliminate all forms of scientific and religious universe-origin prejudices)

6. SCIENCE180 ACADEMY FOR CHILDREN (This Academy breaks down origin key topics into language that children can fully understand). This is the only Science180 Academy that your whole family will like and enjoy together, and which will set children on the path of success by accurately showing them early in life the formation of the universe, and how to detect errors in theories or stories that would misguide them as they grow up.

7. SCIENCE180 ACADEMY OF THE PSEUDEPIGRAPHA AND SPIRITUAL WORLD (Only one ancient blueprint has the reliable power to help you to accurately decrypt the spiritual origin and history of everything in the universe. If you are a believer and want to delve into the prophetic, angelic, and higher order of knowledge based on the spiritual world, then this Science180 Academy is for you. This program is suitable for those who took at least "Science180 Academy of Creationism". Academy of Creationism).

8. SCIENCE180 ACADEMY OF CREATIONISM (Science180 Creationism is a scientific theory spearheaded by the groundbreaking discoveries of Nathanael-Israel Israel, that scientifically explained the origin of the universe, life, and chemicals using turbulence, and that mathematically reconciled science and the Biblical account of creation for the first time in history. Science180 is different from all existing creationist theories known before 2025. Science180 Creationism reconciled science with the Biblical account of creation, including scientifically proving that the Earth was formed on Day 3, while the Moon and the Sun were formed on Day 4 of creation!). As you attend "Science180 Academy of Creationism", you will receive accurate answers to all your questions concerning the creation of the universe).

9. SCIENCE180 ACADEMY FOR FREETHINKERS & ALL ANTI-CREATIONISTS (This Science180 Academy is designed for evolutionists, anti-creationists, and all other types of unbelievers seeking to rationally explore and understand alternative arguments for creation or formation or origin of the universe, life, and chemicals from a fresh, scientific perspective).

10. SCIENCE180 ACADEMY OF LEADERSHIP-(Also called "Science180 Academy for Leaders", this program will enlighten leaders of organizations on how to solve their people problems, process problems, and profit problems related to the origin of the universe, of life, and chemicals according to their domain of expertise). With "Science180 Academy of Leadership", leaders will gain new insights so they can cast new visions and avoid focusing on screwed-up processes, products, and services related to universe-origin initiatives that need to be fixed, faced, or dealt with. Science180 Academy of Leadership will also equip leaders to address process problems related to inefficiency, gaps, missed opportunities, wasted time and efforts, too many steps, bureaucracy, useless layers between organization and customers concerning the innovation, research methodology, research, product

development, strategic planning, workforce diversity in alignment with the historic Science180 breakthroughs so that they can sell more often at full price, avoid regrets in the end, open new markets focusing on real solutions, expand their products and services lines, cut useless costs and research, stop wasting time on useless products that will yield nothing, start focusing on the real money-making problems, ….

11. SCIENCE180 ACADEMY FOR GOVERNMENTAL AGENCIES (Do you want to know how and why most nations and governments are wasting millions of dollars on universe-origin and life-origin researches they don't need … and how to avoid it? Indeed, for most developed nations, and even for some under developed countries, universe-origin projects can cost billions of US dollars and other expensive things that cannot be afforded without sacrificing crucial priorities. Even in developed countries, the impact and the return of investment of the space researches are subject of intense political and economic debates. What if your nation or institution can reduce wasteful spending on universe-origin and life-origin researches, as well as your dependency of wrong theories on the origin of the universe and life? "Science180 Academy for Governmental Agencies" will show you how to use the latest scientific breakthrough to better understand the origin of the universe without wasting money on what is already known or what we think we don't know, but that most scientists ignore. Having spent years accurately decoding the origin of the universe, of life, and of chemicals, Dr. Nathanael-Israel Israel delivers science-backed insight to properly understand all the processes connected to the universe formation—so you don't waste more money and time on trying to research the beginning of the cosmos, but to focus on reducing budget of spatial agencies, focus on real science, cutting-edge research, and things that inevitably lead to discovery and innovation).

12. OTHER SCIENCE180 ACADEMY: If you did not relate with any of the Science180 Academies mentioned above, but you are still interested in learning something specific about the origin of the universe, life, and chemicals that better fits your needs, please visit Science180Academy.com to contact us so we can discuss that with you.

Learn more at www.Science180Academy.com

5.6. Genesis of chemical particles, rings, and crusts of celestial bodies

Chemical particles were not formed by chance but by complex processes that you deserve to understand. In this chapter, you will discover a few major facts that helped me to crack the mysteries behind the formation of chemical particles. In previous chapters, I introduced the processes that gave rise to the precursors of chemical particles. In *"Turbulent Origin of Chemical Particles*," I devoted hundreds of pages to chemistry, but because chemistry may not interest the general public much, I will not dwell on it here. Instead, I will recall just a few aspects of the characteristics of chemical particles to give you a general understanding of how they were formed, hoping that those interested in deepening their knowledge of chemistry can refer to my other books.

When chemical particles are mentioned, people mostly think about atoms, which are the common matter in our everyday life. Hydrogen (annotated by the letter H) and oxygen (annotated by the letter O) are examples of atoms, and, together, they bind to form ordinary water (annotated H_2O, meaning a molecule containing two atoms of hydrogen and one atom of oxygen). However, some chemical particles called subatomic particles are smaller than atoms, and some of them are found inside atoms. For instance, electrons, protons, and neutrons (all of which are found in atoms) are examples of subatomic particles. Because most people are not familiar with subatomic particles, I will not elaborate much on these microscopic particles. Other chemical particles are larger than atoms; some are chemical compounds, minerals, and rocks. Diamond is an example of a mineral, while granite is an example of a rock.

To explain the origin of chemical particles, among many other things, I had to rely on the data available on the abundance of chemical elements and compounds in (1) the universe or in our galaxy, (2) in the Sun, (3) in human beings, (4) in oceans and sea waters, (5) in the Earth's crust, (6) in the atmosphere of the planets in the Solar System, and in the atmosphere and crust of the Moon. The raw data from the website of NASA showed that the chemical composition of the atmosphere of the planets in the Solar System varies from one planet to another. The characteristics of chemical particles (e.g., mass, radius, density) and those of celestial bodies helped me uncover secrets about their formation. Because celestial bodies have different characteristics, by carefully analyzing and comparing their properties with those of the chemical particles they contain, I detected some factors that could have prevailed during the genesis of the chemicals. For instance, some dense chemical elements seem to be also abundant in dense celestial bodies. In contrast, less dense chemicals are generally abundant in less dense celestial bodies.

Just as water evaporating into the atmosphere can condense and birth different kinds of precipitations (e.g., rain, hail, fog, snow, sleet, ice, dew, etc.), and just as a cooking flour can be used to make various kinds of meals, and just as the same soil or concrete can be used to construct different kinds of architectural buildings, so also, although under totally different conditions, the turbulent prima materia (the initial or original matter in the universe) was used to fashion different particles and clusters of particles according to the environmental conditions of their precursors. Just as celestial bodies consist of primary bodies orbited by secondary bodies, atoms also behave almost the same. As I demonstrated for celestial bodies, at some point during their genesis, the precursors of chemical particles could have been organized into tiny layers or pockets of matter. According to the amount of matter involved in the formation of their precursors and how they were organized, some particles ended up being elementary, meaning not composed of any other types, while others are composite, meaning composed of other types of particles. The precursors of the composite particles could have been split-gathered into the particles constituting them. Because some chemical particles were being split-gathered as the systems of matter they belong to (e.g., atoms) were being molded, it is imperative not to view composite particles as having been formed after their elementary or constitutive particles were fashioned. Because of the position of their precursors, some elementary particles were clustered into groups, while others were isolated. The number of particles involved in the clustering defined some characteristics of the resulting particles.

As I was pondering how the chemical particles were formed, I realized that, even today, some biological processes simulate what could have happened to the turbulent prima materia (the original matter in the universe), yielding various chemical elements. Because most people are familiar with DNA, the chemical material that is responsible for most traits in living organisms, I will use it to illustrate something that happened to the precursors of chemical particles. For instance, scientists have shown that DNA consists of strands that can be wrapped, coiled, and supercoiled, yielding different levels of condensation. The coiling of DNA is one of the biological evidences of the folding, supercoiling, winding, twisting, spiraling, and rolling of biological systems. As I carefully investigated the structure of DNA, I felt that the process of coiling DNA testified to what could have happened to the precursors of some particles on a completely different scale and under different conditions, so that they were molded into various bodies. Just as DNA is coiled and supercoiled into nucleosomes, which are wound up into chromatin and into chromosomes in some organisms, so also the gathering together of chemical particles could have involved complex processes that had some similarities with folding, twisting, rolling, winding, tightening, coiling, and supercoiling. I am not saying that the gathering of chemical particles followed the precise process of DNA supercoiling, but I felt we could learn something from DNA supercoiling to enrich our understanding of what may have happened to the turbulent prima materia as it was being molded into particles.

As the precursors of chemical particles were being squeezed, interactions (that some people called forces) appeared between them. I demonstrated that the so-called forces between chemical particles are the consequence of the processes involved in the split-gathering of their precursors. The energy or forces involved in the squeezing of the chemical particles affected their density and size. The portion of space occupied by the precursors of chemical particles and the degree of their tightness or winding around the nucleus affected the radius, density, and other properties of their atoms, as well as the extent of their interactions with other atoms and particles in their environment. As of 2022, about 118 chemical elements are known in the Solar System, and together they combine in various ways to form chemical compounds.

As the atoms were being formed, the position of their subatomic particles with respect to one another changed due to their rearrangement until they were "locked" into interactions that limited their movements. At the subatomic and atomic levels, rearrangements progressively occurred between the precursors of the particles until they were shaped, "locked up," or settled into their "final" configuration defined by their relationships with one another. Some of these relationships are qualified today as bonds. The distance between chemical particles is governed by laws that depend on their size, nature, and other characteristics. Consequently, when scientists try to engineer some compounds or force some particles to stick with others in which they were not used to in nature, significant changes can occur in their constitutions. That is one of the problems faced by chemists and physicists in the so-called proton radius puzzle, which holds that the size of the proton changes depending on the size of the bodies surrounding it. Likewise, the nature and characteristics of minerals and rocks depend on the nature of their precursors. In *Turbulent Origin of Chemical Particles*," I explained how the diversity of celestial bodies relates to what happened to their chemical particles. In the case of the terrestrial planets, for instance, rocks were piled together to form celestial bodies. While the outer portion of some precursors was volcanoes. Some solid rocks were joined together, while others were deposited on top of others. The fluids within and between the precursors also acted as solvents that glued some rocks and their constituents.

Because most of the precursors of their particles did not form very dense particles and because of their huge size, the precursors of stars were unable to solidify their outer surface as a crust. Lacking an outer solid shell, stars are therefore believed to be mostly made of hydrogen and helium. At the planetary level, the precursors of some planets formed dense bodies with crusts that cover their outer parts. That is the case for the terrestrial planets such as Mercury, Venus, Earth, Mars, and Pluto. In contrast, the precursors of some giant planets, such as Jupiter and Saturn, were unable to form many dense particles, which, put together, could have formed a dense crust. Instead, they formed giant gaseous planets. In contrast, the precursors of Uranus and Neptune ended up forming icy planets. Between those planets, asteroids of various chemical compositions were formed. Depending on their position, some comets are believed to have a chemical composition intermediate between the metal-rich terrestrial planets and the volatile-rich outer bodies. Moreover, because of the circumstances of their formation, the main-belt asteroids were not gathered into a single asteroid or planet by piling their rocks, but instead the rocks were scattered and became individual asteroids organized into a belt.

Because it was bigger than the precursor of any other body in the Solar System, the precursor of the Sun was not much compressed or packed by the forces that shaped it into a uniform celestial body. Instead, most of its constituents or particles were less dense and less compressed than the precursors of some other bodies in the Solar System. The precursor of the Sun could have been like a big light or a very hot body filled with energy. The nature of the precursor of the Solar System explains why the center of the Earth and of most planets could contain lava or magma, which can erupt into a volcano. As the precursors of some celestial bodies in the Solar System were moved farther from the Sun, their fluids cooled down because of the decrease in temperature, at least at their surface. This explains why the surfaces of the terrestrial planets (like Earth) are hard or crusty, whereas their interiors may be filled with molten rock, or magma, or fiery materials. All these variations were caused by environmental conditions that affected the nature of the chemicals formed from the precursors of the celestial bodies.

Because of its massive form and huge light, the Sun emits a lot of heat into its surroundings; it never experiences a night, has been very hot, and is not affected by the cold of its surrounding space. The precursors of photons (the particles of light) were matter that were contracted, compacted, and packed together to form big clusters or particles that became photons. Therefore, photons move freely at high speed. Because the particles in the Sun are not tightly tied together, some of them, including photons, are easily released. However, planets and satellites have their constitutive particles highly tightened and do not escape or move out easily. Because the precursors of the particles in the celestial bodies that are not stars were combined with others to form particles more complex than photons, they do not exist as light. The energy in particles other than photons somehow bears witness to their original formation from something like the precursor of light particles. The big planets like Jupiter and Saturn are not like the Sun because, during their formation, their particles were transformed into particles more complex than the photons that form light. Just as the chemical composition of the Earth is not the same everywhere on Earth, so also the chemical composition of the Sun and of all other celestial bodies cannot be expected to be the same in all locations. For instance, although the Sun is dominated by hydrogen and helium, it also contains small amounts of heavy atoms because, when the precursors of the matter in the Sun were being molded into atoms, the environmental conditions in some spots or pockets allowed their gathering into heavy elements instead of the smaller and lighter hydrogen or helium. On Earth, some spots are richer in some minerals than others. Some countries are richer in certain precious materials than others. All these variations were caused by how the Earth's precursor was molded into different chemicals by the ambient conditions that prevailed at specific locations during Earth's formation.

In summary, stars are like naked celestial bodies that do not have a crusty cover like terrestrial planets do. On the other hand, celestial bodies that have a crust are like clothed bodies, whose nakedness (fluids in their interior) is covered by a crust. The lava inside the Earth and other planets shows that their precursors were very hot and cooled to form a crusty outer layer. The lava or magma inside some planets and satellites provides evidence that the precursors of their constituents did not undergo any changes that could have solidified them. But as seen with volcanic activities, lava or magma can solidify. At this point, so I do not annoy those who are not interested in chemistry too much, I will finish this chapter by presenting the name (with the symbol in parentheses), atomic mass, discovery year, discoverer's name, and the origin of the name of the things or persons the chemical element was named after. Although most people are not familiar with chemical elements, I felt like they may appreciate knowing a little more about the variables I presented in the following tables. The origin of the names of the chemical elements is based on the work of Yinon (2016).

To learn more about chemical particles, check out *"Turbulent Origin of Chemical Particles*," the scientific book I wrote on them.

Table 1a: Name, mass, discovery year, discover name and meaning of the names of the chemical elements

Name (Symbol)	Atomic Mass (gram per mol)	Discovery year	Discoverer name	Origin of name (Name after)
Hydrogen (H)	1.008	1766	Henry Cavendish	Greek words hudôr (water) and gennan (generate)
Helium (He)	4.003	1895	Sir William Ramsay and Cleve Per Teodor	Greek word hêlios (sun)
Lithium (Li)	6.941	1817	Johann August Arfvedson	Greek word lithos (stone)
Beryllium (Be)	9.012	1798	Fredrich Wohler and/or Nicholas Louis Vauquelin	The mineral beryl
Boron (B)	10.811	1808	Sir Humphry Davy, J.L Gay-Lussac, and Louis-Joseph	Borax and carbon
Carbon (C)	12.011	1694	Unknown	Latin word carbo for coal and and charcoal
Nitrogen (N)	14.007	1772	Daniel Rutherford	French word "nitrogene"
Oxygen (O)	15.999	1774	Joseph Priestly and & Scheele Carl Wilhelm	Greek words oxus (acid) and gennan (generate)
Fluorine (F)	18.998	1886	Joseph Henri Moissan	Latin word fluo (flow)
Neon (Ne)	20.180	1898	Sir William Ramsay & Travers Morris	Greek word neos (new)
Sodium (Na)	22.990	1807	Sir Humphrey Davy	Soda (Na2CO3)
Magnesium (Mg)	24.305	1755	Sir Humphrey Davy and/or Joseph Black	City of Magnesia in Europe
Aluminum (Al)	26.982	1825	Hans Christian Oersted	Latin word alumen
Silicon (Si)	28.086	1823	Jons Jacob Berzelius	Latin word silex or silicis meaning flint
Phosphorus (P)	30.974	1669	Hennig Brandt	Greek words phôs (light) and phoros (bearer)
Sulfur (S)	32.066	BCE	Unknown	Latin word sulfur (brimstone)
Chlorine (Cl)	35.453	1774	Carl Wilhelm Scheele & Strutt John	Greek word khlôros (green)
Argon (Ar)	39.948	1894	Sir William Ramsay	Greek word argon (inactive)
Potassium (K)	39.098	1807	Sir Humphrey Davy	Potash
Calcium (Ca)	40.078	1808	Sir Humphrey Davy	Latin word calcis (lime)
Scandium (Sc)	44.956	1879	Lars Fredrik Nilson	Scandinavia (which
Titanium (Ti)	47.867	1791	William Gregor & Klaproth Martin Heinrich	Greek word titanos (Titans)
Vanadium (V)	50.942	1801	Del Rio Andrés Manuel & Nils Sefstrom Gabriel	Vanadis (a Scandinavian goddess)
Chromium (Cr)	51.996	1797	Louis Vauquelin	Greek word chrôma (color)
Manganese (Mn)	54.938	1774	Johann Gahn Gottlieb	Latin word mangnes (magnet)
Iron (Fe)	55.845	BCE	Unknown	Latin
Cobalt (Co)	58.933	1735	George Brandt	German word kobalt or kobold (evil spirit)
Nickel (Ni)	58.693	1751	Alex Fredrik Cronstedt	German word kupfernickel (false copper)

Table 1b: Name, mass, discovery year, discover name and meaning of the names of the chemical elements

Name (Symbol)	Atomic Mass (gram per mol)	Discovery year	Discoverer name	Origin of name (Name after)
Copper (Cu)	63.546	BCE	Unknown	Latin word cyprium, after the island of Cyprus (a major mining place)
Zinc (Zn)	65.39	1500	Andreas Marggraf	German word zin (meaning tin)
Gallium (Ga)	69.723	1875	Paul Emile Lecoq de Boisbaudran	Latin word Gallia, the old name of France
Germanium (Ge)	72.61	1886	Clemens Winkler	Latin word Germania, meaning Germany
Arsenic (As)	74.922	1817	Albertus Magnus	Greek word arsenikos and Latin word arsenicum
Selenium (Se)	78.96	1817	Jons Jacob Berzelius	Greek word Selênê (Moon)
Bromine (Br)	79.904	1826	Antoine-Jérôme Balard	Greek word brômos (stench)
Krypton (Kr)	83.8	1898	Sir William Ramsay & Morris Willliam Travers	Greek word kryptos (hidden)
Rubidium (Rb)	85.468	1861	Bunsen Robert Wilhelm & Kirchhoff Gustav Robert	Latin word rubidus (red)
Strontium (Sr)	87.62	1790	Adair Crawford	Strotian (a Scottish town)
Yttrium (Y)	88.906	1794	Johann Gadolin	Ytterby (a town in Sweden)
Zirconium (Zr)	91.224	1789	Martin Heinrich Klaproth	Zircon (a mineral)
Niobium (Nb)	92.906	1801	Charles Hatchet	Niobe, daughter of mythical king (Tantalus)
Molybdenum (Mo)	95.94	1778	Carl Wilhelm Scheele	Greek word molubdos (lead)
Technetium (Tc)	98	1937	Carlo Perrier & Segrè Emilio	Greek word technêtos (artificial)
Ruthenium (Ru)	101.07	1844	Karl Klaus Karlovich	Latin word Ruthenia (Russia)
Rhodium (Rh)	102.9055	1803	William Wollaston Hyde	Greek word rhodon (rose)
Palladium (Pd)	106.42	1803	William Wollaston Hyde	Greek goddess of wisdom (Pallas) & also after an asteroid
Silver (Ag)	107.868	BCE	Unknown	Old English word seolfor (silver)
Cadmium (Cd)	112.411	1817	Fredrich Stromeyer	Greek word kadmeia (ancient name for calamine) & Latin word cadmia
Indium (In)	114.818	1863	Ferdinand Reich & Richter Hieronymus	Indigo color seen in its spectrum
Tin (Sn)	118.71	BCE	Unknown	Latin
Antimony (Sb)	121.76	BCE	Unknown	Greek words anti (opposed) & monos (solitude), hence "not alone"
Tellurium (Te)	127.6	1782	Franz Muller von Reichenstein	Greek word tellus (Earth)
Iodine (I)	126.904	1811	Bernard Courtois	Greek word iôdes (violet)
Xenon (Xe)	131.29	1898	Sir William Ramsay & Travers Morris William	Greek word xenon (stranger)
Cesium (Cs)	132.905	1860	Fustov Kirchoff & Bunsen Robert	Latin word caesius (sky blue)
Barium (Ba)	137.327	1808	Sir Humphrey Davy	Greek word barys (heavy)

Nathanael-Israel Israel: Known as the #1 Universe-Origin, Life-Origin, and Chemicals-Origin Scientist & Mathematician

Table 1c: Name, mass, discovery year, discover name and meaning of the names of the chemical elements

Name (Symbol)	Atomic Mass (gram per mol)	Discovery year	Discoverer name	Origin of name (Name after)
Lanthanum (La)	138.906	1839	Carl Mosander	Greek word lanthaneis (to lie hidden)
Cerium (Ce)	140.116	1803	Hisinger Wilhelm & Berzelius Jöns Jacob	Ceres (a main belt asteroid)
Praseodymium (Pr)	140.908	1885	Baron Auer Von Welsbach	Greek words prasios (green) & didymos (twin)
Neodymium (Nd)	144.24	1925	Baron Auer Von Welsbach	Greek words neos (new) & didymos (twin)
Promethium (Pm)	145	1945	J.A. Marinsky & Glendenin, L. E.	Prometheus, a god who is believed to have stolen fire from the sky and gave it to man
Samarium (Sm)	150.36	1879	Paul Emile Lecoq de Boisbaudran	Smarskite (a mineral)
Europium (Eu)	151.964	1901	Eugene Demarcay	Europe
Gadolinium (Gd)	157.25	1880	Jean de Marignac	Gadolinite
Terbium (Tb)	158.925	1843	Carl Gustav Mosander	Ytterby (a town in Sweden)
Dysprosium (Dy)	162.5	1886	Paul Emile Lecoq de Boisbaudran	Greek word dysprositos (hard to get at)
Holmium (Ho)	164.930	1878	Marc Delafontaine & Soret J. Louis	Latin word Holmia (Stockholm)
Erbium (Er)	167.26	1843	Carl Gustav Mosander	Ytterby (a town in Sweden)
Thulium (Tm)	168.934	1879	Per Theodor Cleve	Thule (ancient name of Scandinavia)
Ytterbium (Yb)	173.04	1878	Jean Charles Galissard de Marignac	Ytterby (a town in Sweden)
Lutetium (Lu)	174.967	1907	Georges Urbain	Lutetia, an ancient name of Paris
Hafnium (Hf)	178.49	1923	Dirk Coster & George Charles De Hevesy	Latin word Hafnia (Copenhagen)
Tantalum (Ta)	180.948	1802	Anders Gustav Ekeberg	King Tantalus (a mythological Greek king)
Tungsten (W)	183.84	1783	Fausto and Juan Jose de Elhuyar	Swedish words tung sten (heavy stone)
Rhenium (Re)	186.207	1925	Walter Noddack	Rhines provinces of Germany
Osmium (Os)	190.23	1803	Smithson Tenant	Greek word osmë (odor)
Iridium (Ir)	192.217	1804	Smithson Tenant	Latin word iridis (rainbow)
Platinum (Pt)	195.078	1735	Julius Scaliger	Spanish word platina (little silver)
Gold (Au)	196.967	BCE	Unknown	Old English word geolo (yellow)
Mercury (Hg)	200.59	BCE	Unknown	Mercury (name of a planet and also of a god)
Thallium (Tl)	204.383	1861	Sir William Crookes	Greek word thallos (young shoot)
Lead (Pb)	207.2	BCE	Unknown	Greek word protos (first)
Bismuth (Bi)	208.980	1400	Georgius Agricola	German word wissmuth (white mass)
Polonium (Po)	209	1898	Pierre and Marie Curie	Poland

Table 1d: Name, mass, discovery year, discover name and meaning of the names of the chemical elements

Name (Symbol)	Atomic Mass (gram per mol)	Discovery year	Discoverer name	Origin of name (Name after)
Astatine (At)	210	1940	Dale R. Corson & K. R. Mackenzie	Greek word astatos (unstable)
Radon (Rn)	222	1898	Fredrich Ernst Dorn	Radium
Francium (Fr)	223	1939	Marguerite Derey	France
Radium (Ra)	226	1898	Pierre and Marie Curie	Latin word radius (ray)
Actinium (Ac)	227	1899	Andre Debierne	Greek word aktinos (ray)
Thorium (Th)	232.038	1828	Jons Berzelius	Thor (a Scandinavian god)
Protactinium (Pa)	231.036	1917	Fredrich Soddy, John Cranston & Otto Hahn	Greek word protos (first)
Uranium (U)	238.029	1789	Martin Heinrich Klaproth	Uranus, a planet
Neptunium (Np)	237	1940	Edwin M. McMillan & Philip H. Abelson	Neptune, a planet
Plutonium (Pu)	244	1940	Glenn T. Seaborg	Pluto, a planet
Americium (Am)	243	1945	Glenn T. Seaborg	America
Curium (Cm)	247	1944	Glenn T. Seaborg	Pierre and Marie Curie (pioneers of radioactivity
Berkelium (Bk)	247	1949	Glenn T. Seaborg	Berkeley, a city in California
Californium (Cf)	251	1950	Glenn T. Seaborg	California (State and University)
Einsteinium (Es)	252	1952	Glenn T. Seaborg; Argonne, Los Alamos, University of California	Albert Einstein (the author of the relativity theory)
Fermium (Fm)	257	1953	Glenn T. Seaborg; Argonne, Los Alamos, University of California	Enrico Fermi (a famous scientist after whom the subatomic particles called fermions are named)
Mendelevium (Md)	258	1955	Glenn T. Seaborg	Dmitri Ivanovitch Mendeleyev, the chemist who pioneered the chemical elements table named after him.
Nobelium (No)	259	1957	Nobel Institute for Physics	Alfred Nobel, the one who founded the Nobel Prize awards
Lawrencium (Lr)	262	1961	Albert Ghiorso	Ernest Lawrence (a physicist)
Rutherfordium (Rf)	267	1969	Albert Ghiorso	Lord Rutherford, a New Zealand chemist and physicist
Dubnium (Db)	268	1970	Albert Ghiorso	Dubna, Russia

Table 1e: Name, mass, discovery year, discover name and meaning of the names of the chemical elements

Name (Symbol)	Atomic Mass (gram per mol)	Discovery year	Discoverer name	Origin of name (Name after)
Seaborgium (Sg)	271	1974	Albert Ghiorso and/or Glenn T. Seaborg	Glenn T. Seaborg (the one who discovered many transuranium elements)
Bohrium (Bh)	272	1976	Peter Armbruster, Gottfried Munzenber and others; Oganessian	Niels Bohr (Danish physicist)
Hassium (Hs)	270	1984	Peter Armbruster, Gottfried Munzenber and others	Latin word Hassias, a German state.
Meitnerium (Mt)	276	1982	Heavy Ion Research Laboratory; Armbruster Paula & Dr. Gottfried Muenzenberg	Lise Meitner (an Austrian physicist)
Darmstadtium (Ds)	281	1987	Organessian, et al.; Armbruster, Paula & Muenzenberg, Dr. Gottfried	Darmstadt, Germany, where the element was first synthesised
Roentgenium (Rg)	280	1994	Sigurd Hofmann, Armbruster, Paula & Muenzenberg, Dr. Gottfried	Wilhelm Conrad Röntgen (a German physicist)
Copernicium (Cn)	285	1996	S. Hofmann, V. Ninov, F. P. Hessbuger	Nicolaus Copernicus, Polish astronomer
Nihonium (Nh)	286	2004	Riken (a Japanese Institute of Physical and Chemical Research)	Japanese city Nihon, a place in Japan where the element was first synthesized
Flerovium (Fl)	289	1998	Joint Institute for Nuclear Research (JINR) and Lawrence Livermore National Laboratory(LLNL)	Flerov Laboratory of Nuclear Reactions of the Joint Institute for Nuclear Research in Dubna, Russia, where the element was synthesised; the lab itself was named after Georgy Flyorov, a Russian physicist
Moscovium (Mc)	290	2003	Joint Institute for Nuclear Research (JINR) and Lawrence Livermore National Laboratory(LLNL)	Moscow Oblast, Russia, where the element was first synthesised
Livermorium (Lv)	293	2000	Joint Institute for Nuclear Research and Lawrence Livermore National Laboratory	City of Livermore in California where is based the Lawrence Livermore National Laboratory, which collaborated with the Joint Institute for Nuclear Research in Dubna, Russia, to discover livermorium.
Tennessine (Ts)	294	2009	Joint Institute for Nuclear Research, Lawrence Livermore National Laboratory, Vanderbilt University, and Oak Ridge National Laboratory	Tennessee (a state in the USA, where Oak Ridge National Laboratory is located)
Oganesson (Og)	294	2002	Joint Institute for Nuclear Research and Lawrence Livermore National Laboratory	Yuri Oganessian (a Russian-Armenian nuclear physicist)

Another Book by Nathanael-Israel Israel:
TURBULENT ORIGIN OF CHEMICAL PARTICLES

FIND ALL THE RELIABLE, CONVINCING, SCIENTIFIC ANSWERS YOU NEED TO SUCCESSFULLY DECODE THE ORIGIN OF CHEMICAL PARTICLES SAFELY

Where did all elementary particles and composite particles including atoms, molecules, minerals, and rocks come from? What are the fundamental factors, the machinery, and the generic processes that defined their formation and proprieties? What was the nature of their precursors at the beginning of the universe and what underlying processes shaped or molded them into the chemicals we know today? What was the primary cause of the abundance and diversity of chemicals in the celestial bodies in the universe? What is the accurate link between the formation of chemical particles and the formation of galaxies, stars, planets, asteroids, and satellites? What light can the origin of chemicals shed on the real cause and meaning of gravity and the other so-called fundamental forces in nature? How does the formation of the chemical particles fit into the big picture of the formation of the universe?

After studying these questions for more than 12 years, Dr. Nathanael-Israel Israel discovered that the proper understanding of the origin of chemical particles is a very challenging but profitable task that requires original, scientific, mathematic, and philosophic efforts beyond the current state of modern science—until recently. The solution for all of these puzzling problems: *"Turbulent Origin of Chemical Particles"*, the straightforward and trustworthy book that will help you to quickly, cheaply, easily, and efficiently navigate everything you need to know to finally solve the hard problems about the origin, the formation, and the functioning of all chemical particles. Whether you are a chemist, a biochemist, any other scientist, an engineer, as long as you have a reasonable background in chemistry but ignore how to scientifically demonstrate the origin of all chemical particles, this marvelous book is for you! Amazingly packed with eye-popping analysis, fantastic graphs, tables, and the historic formula that broke the universe-origin code, *"Turbulent Origin of Chemical Particles"* will:

- Make it easier than ever for you to properly understand, decrypt, and articulate the real origin of natural chemical particles in the universe, therefore freeing you from false and boring explanations of the origin of all matters, and embrace the proven theory that opens doors to unparallel opportunities

- Professionally teach you how to transform the true knowledge of the origin of chemical particles into insights that significantly add value to your life in less time, and successfully establish you as a symbol of freedom, power, creativity, and originality in your field of expertise

- Fire you up to become the best version of you, and to cause positive changes to your initiatives that will profit you nonstop
- Discover thrilling illustrations and unconventional explanations of the formation of all matter in the universe, written in a simple language that brings humankind much closer to the complete deciphering of the mysteries at the very heart of chemistry, and open the way to a future of technology, innovation, discoveries, and breakthroughs
- Equip you to bypass technical knowledge that restricts non-experts from accessing the origin-related secrets contained in the massive scientific data, and get to the bottom of origin-related mysteries regardless of your background so you can empower yourself to leave unforgettable marks in your field of expertise
- Learn more at Science180.com/chemical

With *"Turbulent Origin of Chemical Particles"*, the accurate decrypting and understanding of the formation of chemicals has never been profitable and easy. Hence this great book is THE ultimate how-to guide for great people wanting to correctly decode the origin of the chemicals and positively transform their lives. Get this celebrated book today. Don't wait!

Known as the nonconformist, rule-breaker, and accurate demonstrator of the universe-origin, **Dr. Nathanael-Israel Israel** is the founder of Science180, the one-stop for answering the most crucial universe and life's origin questions. He has had the honor to be acknowledged as the fearless universe-origin decryption trailblazer. Learn more at Israel120.com.

5.7. Stunning details about the origin of planetary rings

As I handled chemical particles in this segment, I felt like it was appropriate to say a few words about the formation of rings. Indeed, rings are solid materials (e.g., dust, moonlets, and other small objects) found in the form of a disc around an astronomical object. When rings are considered, people usually think about the rings of Saturn. That is why, in most pictures of the Solar System that show the planets, Saturn is usually the only planet shown with rings, as its rings are the largest. However, rings are found around all the 4 giant planets, and recent scientific evidence suggests that they are also around other celestial bodies, including satellites, asteroids, and terrestrial planets (e.g., Mercury, Venus, and the Earth). Some of these rings can be seen with telescopes on Earth and are estimated to be millions of miles wide. Some satellites, such as Triton, the largest moon of Neptune, have their own rings. More than 48 rings have been observed around the 4 giant planets in the Solar System. Some rings are located between the primary planet and the innermost satellite in their planetary system, meaning that some planets have no satellite near their innermost rings, while others have satellites embedded in their rings.

The size, nature, and chemical composition of the particles and rocks in the rings depend on the types of rings. In general, the particles in the rings are not larger than 20 meters. Some rings are faint, while others are very dense. Considering the properties of the rings and those of the celestial bodies containing them, I demonstrated that the origin of the rings is connected to the turbulence that formed the planetary systems they belong to. I showed that, as the fluid layers in the precursor of the secondary bodies in the planetary systems started going through turbulence, they sometimes "failed" to gather together all of their particles into a single body and to form a single satellite that has no trail of dust around it. Instead, some particles were gathered together into rocks of various sizes, while others were gathered together into small particles of various sizes. The particles that were gathered into larger bodies became the satellites, while those that were not incorporated into the precursor bodies formed the rings. Expressed in a different way, because the particles in some fluid layers of precursors of secondary bodies could not be gathered together into satellites, they were scattered into large particles and/or rocks, which, from far, appear as rings. The fluid layers that were not gathered together at all, and the leftover particles from the layers that were gathered, formed the particles and rocks present in the rings. In other words, during the formation of the rings and satellites, there was a time when the precursors of the rings and satellites were particles that were split-gathered into the precursors of the rings and the precursors of the satellites as their turbulence was underway. As turbulence was developing and fluids were moving, mixing, rotating, spiraling, and packing, some fluids were packed into bigger clusters that yielded the satellites, while other fluids did not rotate enough to be amassed into bigger bodies but into smaller particles. spread over the space occupied by their precursors. The particles that were not assembled into satellites became the constituents of the rings. In contrast, the gathering together of the precursors of particles into denser bodies led to the formation of the satellites. Hence, some satellites are embedded inside rings. As the rings were forming, the satellites and the primary planet in their planetary system were also undergoing the processes that had given rise to them. By the time the primary planet, its satellites, and its rings had formed, they were organized into a planetary system as sketched below.

Some rings are faint because parts of the precursors of the particles that were present in their neighborhood were gathered together into satellites embedded in those rings. Some of the empty spaces or gaps between rings can be explained by the demarcation of the space between the fluid layers of their precursors and the incorporation of the precursors of some particles into satellites. The more the particles in the precursors of the secondary bodies were gathered together into satellites, the less material was left to form the rings, hence a gap or a faint ring between some satellites. The density and brightness of the rings could have been defined by the number of particles in the precursors of bodies that were not gathered together into the satellites but were left out to become part of the rings. The fewer the particles in the rings, the brighter it may look, while the more particles in the rings, the denser it would be. I showed that, due to the nature of the instability and turbulence it underwent, the precursor of the Saturnian rings, which could have been used to form other Saturnian satellites or to increase their size, was instead spread out to form the prominent Saturnian rings. As far as density is concerned, the bodies in the Saturnian planetary system are usually the least dense. Saturn itself is the least dense planet in the Solar System. Because the precursors of the Saturnian satellites could have been less dense, the precursors of the Saturnian rings could have diffused more and/or were unable to amass into bigger rocks and bigger satellites.

As a take-home message, the particles in the rings were not transported to their current location by other celestial bodies (as some theories assumed) but were born from the precursors of the secondary bodies in their planetary systems. In the areas where rings are found, small rocks or clusters of particles were molded but probably could not be amassed into larger celestial bodies, such as satellites, or incorporated into their neighboring satellites. As the precursors of the rings could not collect into bigger bodies, they were spread all over the domain where they were formed. The color, density, brightness, thickness, and other characteristics of the rings depend on their precursors and how their environmental conditions shaped them. The variation of the density, faintness, or brightness of the rings can be explained by the variation of the proportion of the precursors of the secondary bodies gathered together into satellites or left over to be incorporated into rings, and how this leftover was scattered throughout. To learn more about chemical particles, please visit the book I wrote on them.

'Science180 Academy' Success Strategy
SCIENCE180 MASTER CLASS

Hear the greatest scientific and philosophic lessons from top scientists, philosophers, thinkers, and public figures who have realized historic mistakes they made in life (concerning the origin of the universe, life, and chemicals), and that they corrected thanks to the historic discovery of Nathanael-Israel Israel, the world's first 180Scientist who founded Science180 and who is known as the one who truly decrypted the universe-origin for the first time. In their own words, these renowned personalities share with the world key lessons they have learned in life and how people can learn from their experiences to improve lives instead of repeating their mistakes that many people still ignore at their own perils. To learn more, contact us at Science180.com/contact.

5.8. Everyone stopped for THIS secret of the formation of galaxies

You have already heard about constellations, galaxies, and stars, but you probably did not know how they were formed. Here, in a few words, I will help you to understand just that. Indeed, the night sky is filled with bright objects, including stars, planets, satellites, and asteroids, all of which fascinate people of all ages. However, not every bright object in the sky is a star. For instance, celestial bodies that were once believed to be stars have been proven to be planets. For instance, because they are very bright in the night sky, Venus and Mercury were considered stars for a long time before being proven to be planets. Likewise, a few centuries ago, people considered Jupiter and Saturn as stars, yet they are planets. People who are not versed in astronomy can still mistakenly think that all the planets I just mentioned are really stars.

In ancient times, stars were grouped into constellations, which are defined as imaginary outlines or patterns of the celestial bodies in the sky, symbolizing animals, mythological entities (e.g., individuals, gods, or idols), and other things. The definition of a constellation varies across nations and cultures worldwide. As of 2020, the International Astronomical Union recognized about 88 constellations, of which about 60% are predominant in the southern sky.

With the advancement of science, the stars in the sky have been organized into galaxies, which, in short, are clusters of stars apparently "bound together." According to NASA, a *"galaxy is a gravitationally bound system of stars, stellar remnants, interstellar gas, dust, and dark matter"*. However, based on my investigation into the origin of the universe, as I will elaborate in the next chapter, I do not think that gravity binds the stars in a galaxy. Because of the vast distances between stars, a unit like miles or kilometers is not adequate to express the distance between them. In contrast, a unit called a light year (the distance traversed by light in one year at the speed of about 300,000 km/s) is usually considered. Because stars, planetary systems, planets, asteroids, satellites, and chemical particles are constituents of galaxies, most of the things I already said about their origin happened inside the galaxy they belong to. In *"Turbulent Origin of the Universe,"* I spent tens of pages talking about galaxies and stars, but here, I will just present a few data points about how they were formed.

The size of the stars can range anywhere from a few kilometers to millions of kilometers. For instance, the star called Betelgeuse is alleged to have a diameter of 950 times that of the Sun. At the beginning of the 21st century, scientists said that the observable universe contains hundreds of billions of galaxies. But as of today, the observable universe is believed to contain trillions of galaxies. To put the magnitude of that number into simpler terms, the experts said that there are more stars in the universe than all the grains of sand on planet Earth. While efforts have been made to catalog the stars in galaxies, much remains unknown. Examples of "well-cataloged" galaxies are:

- Milky Way Galaxy (the local galaxy that the Solar System belongs to)
- Andromeda Galaxy, the northern sky's brightest galaxy, is believed to be the nearest large galaxy to the Milky Way Galaxy

Some galaxies are believed to be millions of light-years wide. The Andromeda Galaxy and the Milky Way are believed to be separated by millions of light-years. Some people tend to use the distance separating the galaxies to prove that the universe must have been formed millions of years ago, but they fail to understand that the universe has been expanding since its formation and that the initial matter in the universe appeared over a very large space. Hence, the huge distances between the celestial bodies cannot be used to estimate the age of the universe.

While some galaxies, called solitary galaxies, are "isolated," most galaxies are organized into structures called groups, clusters, superclusters, and clusters of superclusters. Clusters of galaxies are said to consist of a few hundred to thousands of galaxies. Superclusters of galaxies are alleged to have thousands of galaxies. Some studies suggest that some superclusters of galaxies can also be organized into even larger superclusters called clusters of superclusters. Seen from a supercluster scale, galaxies look like sheets and filaments surrounding vast empty voids. In other words, gigantic voids are found between the clusters of galaxies and all other types of clusters of stars, just as voids are also found between microscopic particles. Hence, when we look at the sky, we see more voids (the dark spaces between the stars and the planets) than celestial bodies.

Galaxies are usually organized into three components: the spheroidal component, the disk component, and the spiral arms. Fluid layers also existed in the precursors of galaxies, as well as in clusters and superclusters of galaxies. The original mysterious scattering that birthed the precursors of galaxies and their clusters could have moved them very fast, but because of the astronomical distance separating the galaxies and us and because of the difficulty of properly measuring the speed and motion of galaxies in our lifespan, it may be very difficult to prove some details about the motion of galactic clusters and superclusters. However, the squashing of the fluid layers of the precursors of some galaxies could have contributed to inclining, flipping over, or overturning the orbital plane and rotational axis of their daughter bodies. For instance, because of our position within the Solar System and the fact that we are inside the Milky Way, we cannot see the Milky Way properly, only a line of stars. However, the shape of the Milky Way Galaxy can be seen with the naked eye as an arc, suggesting that its precursor may have been trying to wind up or swirl, or may have been thrown and behaved like a projectile. This is like how a softball pitcher throws a pitch to the batter, and as the ball leaves the pitcher's hand, it arches and lowers back down before the batter at home plate. When I looked at the pictures of some galaxies, I noticed that the arms of the spirals are pitched differently. As I considered the data that I analyzed about the celestial bodies and the chemical particles, I understood that the processes that explain the formation and evolution of a vortex (structures usually formed in turbulent fluids) can also (on the astronomical scales) explain the formation of galaxies. I observed many similarities among the configurations of galaxies, vortices, the Solar System, and its planetary systems.

Just as a fluid filament can curl to form a vortex, so also the precursors of galaxies could have initially been astronomical fluid layers, which were curled or wrapped into astronomical 3D bodies called galaxies. The precursors of some galaxies could have been tightly collected into spherical bodies, forming, for instance, elliptical galaxies. The fluid layers of the precursors of other galaxies were unable to compact into elliptical galaxies but instead formed spiral galaxies. The central part, or the core, or the spheroidal component of spiral galaxies was being formed, stars, planets, satellites, asteroids, and microscopic particles were also being formed inside of them. Just as stars are orbited by planets and asteroids, so also the core of the spherical components of galaxies is "orbited" by stars and stellar systems. Just as rings are found orbiting some planets, so also on the galactic scale, the galactic disc contains many stars and planets embedded in it. The precursors of the galactic spirals could have been ejected from the spherical component as the precursors of galaxies were split-gathering.

According to their visual morphology, galaxies are classified into three main types: elliptical, spiral, and irregular. Just as some planets do not have satellites, while others have many, so also galaxies do not have the same components. Some galaxies have more components than others. Spiral galaxies that do not have many stars in their arms are like planets that have fewer satellites. In contrast, spiral galaxies, which have long arms of stars, are like planets with many satellites. In *"Turbulent Origin of the Universe,"* I detailed the developmental stages of the galaxies.

In the early days of my investigation of the origin of the universe, some of the first images I looked at, and which inspired me, were those of hurricanes, cyclones, and whirlwinds. For instance, when I carefully examined the description of cyclones, I noticed significant similarities between the eye of a tropical cyclone and other vortical systems in nature, including tornadoes, waterspouts, and whirlpools. When I compared images of hurricanes to those of the Milky Way Galaxy, I noticed some strong similarities. In those days, I realized that if I could understand the movement of waters in hurricanes, I could better explain the origin of the planets. Having not found much information about the equation of the movement of water in hurricanes, I turned to the kinematics of projectiles. By doing so in 2013, I started having an idea about how the precursors of the satellites could have been positioned at different locations in the fluids of their mother precursors. By November 18, 2013, I had a first impression of how the precursors of the planets could have been formed and progressively split from one another. I would like to recall a note (written in French, as the inspiration came to me in French, not in English) that I made in 2013 in Fig. 20 about how the precursors of the planets could have been placed into orbit. As a souvenir from my 2013 notebook, this sketch gave me an early perspective on the split-gathering of bodies in the universe. As I mentioned in that graph (Fig. 20), it is important not to neglect strange ideas; else, I would have thrown away a lot of ideas I did not initially understand, but which ended up being the bedrock of the layers explaining the origin of the universe.

Fig. 20: Graph from my notebook (November 18, 2013) about how an initial "wind" pushed the precursors of planets into movement

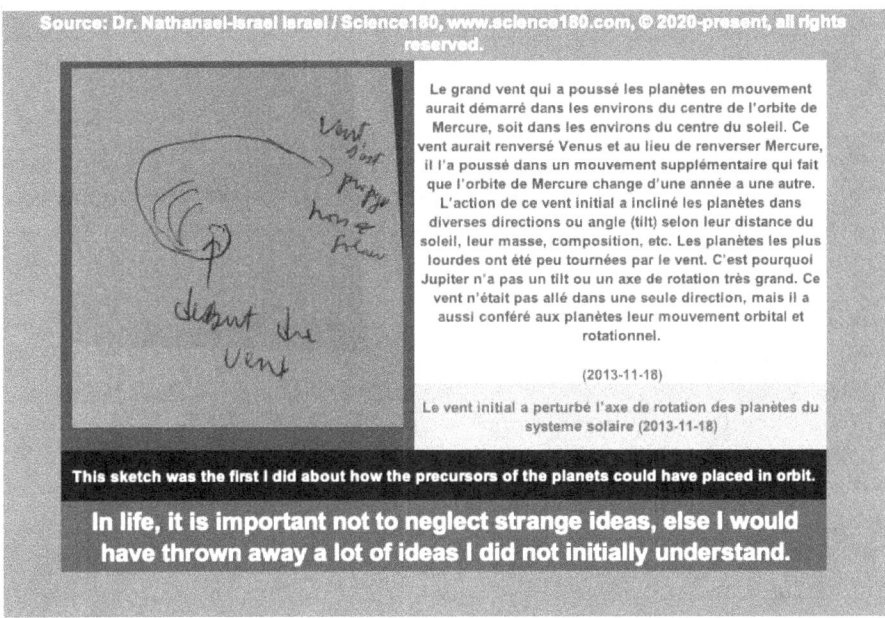

Le grand vent qui a poussé les planètes en mouvement aurait démarré dans les environs du centre de l'orbite de Mercure, soit dans les environs du centre du soleil. Ce vent aurait renversé Venus et au lieu de renverser Mercure, il l'a poussé dans un mouvement supplémentaire qui fait que l'orbite de Mercure change d'une année a une autre. L'action de ce vent initial a incliné les planètes dans diverses directions ou angle (tilt) selon leur distance du soleil, leur masse, composition, etc. Les planètes les plus lourdes ont été peu tournées par le vent. C'est pourquoi Jupiter n'a pas un tilt ou un axe de rotation très grand. Ce vent n'était pas allé dans une seule direction, mais il a aussi conféré aux planètes leur mouvement orbital et rotationnel.

(2013-11-18)

Le vent initial a perturbé l'axe de rotation des planètes du systeme solaire (2013-11-18)

This sketch was the first I did about how the precursors of the planets could have placed in orbit.

In life, it is important not to neglect strange ideas, else I would have thrown away a lot of ideas I did not initially understand.

5.9. Take -home message

In this chapter, I walked you through my thought process for calculating the duration of the universe's formation. I demonstrated that by 26.046 hours after the beginning, fluid layers of the precursor, the bodies orbiting the Sun, started splitting, and by this time, the precursor of the Sun was also formed. The precursor of Mercury was among the first to split from the stack of fluids of the bodies orbiting the Sun. Then, many other fluid layers split from the stack of fluid layers of the bodies orbiting the Sun until the turn of the precursor of the Earth-Moon system came about 67.286 hours after the beginning. Within a few minutes, the precursor of the Earth-Moon system split into the precursors of the Earth and the precursor of the Moon. About 22.44 minutes later, the precursor of the Earth was collected into the Earth, meaning that the Earth was formed 67.659 hours (i.e., 2.819 days) after the beginning. The precursor of the Moon traveled for about 9.546 hours (i.e., semi-major axis of the Moon / escape velocity of the Earth) before being in position to swirl and form the Moon. That swirling occurred within 2.963 hours, meaning that 79.795 hours (i.e., 3.325 days) after the beginning, the Moon was formed. As for the Sun, its precursor was formed about 26.046 hours after the beginning of the formation of the Solar System. That time was required for the fluid of all the bodies orbiting the Sun to escape the precursor of the Sun.

Once the precursor of the Sun was formed, it swirled for about 62.584 hours to form the Sun, meaning that the Sun was formed 88.63 hours (i.e., 26.046 hours +

62.584 hours) or 3.693 days after the beginning of the Solar System. These data confirmed that the Earth was formed on the 3rd day, while the Moon and the Sun were formed on the 4th day after the beginning of the Solar System's formation, which was near the beginning of the universe's formation. Therefore, the scientific evidence also shows that the Earth was formed on the 3rd day after the beginning of the universe, while the Moon and the Sun were formed on the 4th day. This scientific evidence confirms the Biblical story of creation. According to the Bible, the Earth was created on the 3rd day, while the Moon and the Sun were formed on the 4th day. The Bible also explains key turbulent processes related to the formation of these celestial bodies. Just as the Bible says on the first day of creation, the precursors of the bodies were formed; then, fluid layers were formed and separated. The fluid layers began to separate no later than the 2nd day. The Bible also said that the formation of the Earth was completed on the 3rd day, while the formation of the Moon and the Sun was completed on the 4th day.

I also showed that the scientific evidence confirmed that the Biblical creation days were 24 hours each, NOT millions of years as some people think. This scientific evidence supports the Genesis account, which holds that God is the Creator of the universe. The story about the creation of the Earth, the Moon, and the Sun cannot be accurate, and the part that claims God is the creator is false. Considering other things revealed in the book of Moses (the author of the Bible's Book of Genesis), it is important that you reflect on the significance of this chapter seriously. The decision we make regarding the Creator of the universe will be consequential. Therefore, I urge you to reconsider your perspective on the origin of the universe. I searched the literature and realized that, among all the religious and scientific ideologies, none has ever presented scientific, accurate evidence for the formation of the Earth, the Moon, and the Sun as the Bible did. In short, among all the religious narratives of the universe's creation, only the Judeo-Christian story recounted in the Bible is 100% backed by scientific data.

Because my explanation of the Genesis narrative is connected to turbulence, is completely new, and is different from previous interpretations of creation (even by most believers), and because I need to differentiate my creation perspective from all previous theories of creation, I called it "Science180 Creationism." Finally, because I am the one who, for the first time in history, proved the timeline of the universe's formation, I understand that some people may also expect me to demonstrate the universe's age. Therefore, I also addressed this issue in another book. However, due to the scope of this book, I cannot address the age and fate of the universe, as they involve philosophical facts I cannot address here but will in another book. I also did not want to dilute that demonstration here for the sake of pleasing some people, for this book is oriented toward the general public of all backgrounds. For the first time in history, this demonstration settled the disagreement between science and the Biblical story of creation. Remember to check out my other books and see my publications at www.Science180.com.

'Science180 Academy' Success Strategy:
SCIENCE180 BOOKS THAT WILL HELP YOU!

I, Nathanael-Israel Israel, broke down my discovery about the formation of the universe into many books so that you, the readers, can pick the ones that correspond to your needs and interests without disappointing you or wasting your precious time. These books come in many versions (e.g. scientific version, public version, chemical version, biological version, biblical or prophetic version, pseudepigraphic version, and a children's version) targeting people according to their expertise, educational background, and interests as briefed below:

1. **"TURBULENT ORIGIN OF THE UNIVERSE"** (This is the scientific version of my book tailored to scientists and anyone interested in the detailed scientific demonstration of the universe formation). In this book I used the "mother of all turbulences" to scientifically demonstrate the formation of the universe so that scientists can understand and reorient the course of their research, teaching, and publishing, and accept the truth to better live today and forever. Get *"Turbulent Origin of the Universe"* today to begin an incredible journey of accurately decoding the universe and change your life forever! Learn more at Science180.com/scientific

2. **"RECONCILING SCIENCE AND CREATION ACCURATELY"** (this is the book that I called the "Biblical or prophetic version of my book on the universe-origin, and it targets Christians and anyone interested in knowing the Biblical perspective of the creation of the universe). This important book accurately demonstrates the marvelous creation and formation of the universe by God in six consecutive 24-hour-days, and answers many questions about the universe creation, so that after acknowledging Him (who deserves all the glory now and forever), human beings can choose life and avoid the terrible judgment awaiting the unbelievers in the world to come. Get this thoughtful book now to figure out what happened at the beginning, what is coming up, and why it is time to urgently rethink everything you have been told about the universe-origin so you don't eventually regret! Don't say I did not tell you! Learn more at Science180.com/biblical

3. **"TURBULENT ORIGIN OF CHEMICAL PARTICLES"** (Called the "chemical version" of my book on the universe-origin, this elegant book targets chemists, biochemists, and anyone interested in chemistry). With *"Turbulent Origin of Chemical Particles"*, the accurate

decrypting and understanding of the formation of chemicals has never been profitable and easy. Hence this great book is THE ultimate how-to guide for great people wanting to correctly decode the origin of the chemicals and positively transform their lives. Get this celebrated book today. Learn more at Science180.com/chemical

4. *"ORIGIN OF THE SPIRITUAL WORLD"* (This book is what I called the pseudepigraphic or hidden version of my books on the universe-origin, and it is meant for believers who want to tap into a higher level of scriptural secrets that most people may not believe). This book draws the attention of the world toward the pseudepigrapha (a collection of hidden and rejected books, yet filled with deep secrets still valuable today) and explaining how, since thousands of years, God has already revealed deep details about the supernatural origin of the universe, but people (including those who believe or claim to believe in Him) have just refused to literally accept God's mysterious story of creation, which can never be understood by just sticking with conventional science. If you believe in God, have some origin-related questions which answers you cannot find anywhere, not even in the Bible, and if you want to tap into historically neglected revelations to answer fundamental universe and life questions, then be sure to get a copy of *"Origin of the Spiritual World"* today. Learn more at Science180.com/pseudepigrapha

5. *"FROM SCIENCE TO BIBLE'S CONCLUSIONS"* (I called this book the "public version" of my book on the origin of the universe and it is tailored for the general public, and it is a great summary of the scientific version from a perspective that laypeople will fully understand). In this book, I, Nathanael-Israel Israel, broke down the complicated (scientific, philosophical including religious) data about the origin of the universe in a simple language that the general public can fully understand, and know in order to live happily forever. Quickly grab and read this scientifically verifiable, bestselling book to finally get the accurate, jaw-dropping answer that has been rationally shaking both believers, skeptics, and all freethinkers. Don't wait! Learn more at Science180.com/public

6. *"TURBULENT ORIGIN OF LIFE"* (This is the biological or life version of my book on the origin of the universe). It is meant to suit both scientists, nonscientists, and all kids of laypeople, and it decodes the origin of all forms of life so human beings can understand and

and better live.

As of 2025, this book is my only book devoted to the origin of all forms of life, and it will help you to grasp in a simple language what is needed to fully understand the formation of all forms of life. Whether you are a scientist or a layperson, a believer, or a skeptic, you cannot afford to ignore the greater, better, faster, simpler, cheaper, easier, and accurate formula unlocked in this important book that successfully decoded the origin of life. Get *"Turbulence Origin of Life"* today and change lives. Don't wait. Learn more at Science180.com/life

7. **"HOW BABY UNIVERSE WAS BORN"** (How was the universe formed? Did God really form it like some people believe, or did it come out of some long processes? How can we scientifically prove and break down this difficult mystery in a language that children will fully understand and like?) Get the answers as you read this book that I called the "children version" of my book on the origin of the universe and life. Accurately explaining the complex formation of the universe and of life to children can be very hard in our modern world, but by getting *"How Baby Universe was Born"*, you will know the proven formula to help children to easily understand their huge universe-origin and life-origin questions with confidence, humor, and joy. They will surely belly laugh and thank you for it! It is time to buy this pragmatic book and offer it to the children in your life today. Learn more at Science180.com/children

8. **"HOW GOD CREATED BABY UNIVERSE"**. The most difficult part of writing scientific things to children is how to break down complex technical concepts into simple words that they and even anyone who can read and clearly understand (without losing the accurate details and facts). When the topic to address is about the origin of the universe, the task is even more challenging for most people, but not for Nathanael-Israel Israel. As long as you can read, you will find this amazing book extremely helpful to grasp all complicated concepts needed to properly crack the origin of the universe in a language that even children ages 7-12 and anyone who did not go very far in school can fully comprehend.

9. **"SCIENCE180 ACCURATE SCIENTIFIC PROOF OF GOD"** (Whether you are a believer, an unbeliever, a freethinker, administrator, politician, curriculum designer, curriculum specialist, education policymaker, librarian, school board member, parent, researcher, student, teacher, clergy, or a layperson, as long as you are really

seeking to scientifically understand the rational proof of the existence of God, "*Science180 Accurate Scientific Proof of God*" is the much-admired book written for great people just like you). As long as you are interested in the first and the only scientific book that talks to anti-creationists, evolutionists, big bang proponents, atheists, and all other freethinkers and rationalists about the universe formation and they bigly beg to know more about God, the creator, that they mistakenly deny; then this book is for you. As long as you are really seeking to scientifically understand the rational proof of the existence of God, "*Science180 Accurate Scientific Proof of God*" is the much-admired book written for great people just like you. Grab it today and start reading it. Don't wait any longer! Learn more at Science180.com/godproof

If you want to have the entire big picture of my discovery of the origin of the universe, life, and chemicals, and to enlighten your life and career, then plan to get all or some of these books that best suit your needs and interests. For more details, visit Science180.com/books

CHAPTER 6

CAN ANY MODERN SCIENTIFIC THEORY PROVE ALBERT EINSTEIN AND ISAAC NEWTON'S MODELS OF GRAVITATION AND COSMOLOGY WRONG BY PAYING ATTENTION TO ANY CRITICAL DATA THESE INFLUENTIAL SCIENTISTS IGNORED?

Things in nature seem to be connected to one another through interactions more than we can imagine, and some of these interactions are claimed to be mediated by forces. Because some of those so-called forces are essential and appear to constitute the foundation of the connections between bodies in the universe, they are qualified as fundamental. Among these so-called fundamental forces in nature, gravity is the one that the public has heard of the most. Therefore, I will start this chapter by first reviewing what is currently known about the gravity of the celestial bodies in the Solar System. Then, I will expound on other forces in nature.

6.1. Gravity of celestial bodies

Gravity is believed to be responsible for the free fall of objects launched from the Earth. For instance, when you throw an object in the air, gravity is what makes it fall. Among the many variables in physics, gravity is central to many scientific theories, particularly those related to astronomy and cosmology. For instance, gravitational "attraction," gravitational pull, or gravitational force are variables that some scientists use to explain galaxies and stars. The understanding of gravity is one of the main mysteries that scientists have been trying to unravel for centuries. The two main schools of thought about the theory of gravity are:

- The theory of Isaac Newton (published in 1686) was based on a hypothetical "action at a distance" and

- The theory of Albert Einstein (published in 1915) is based on a hypothetical "curvature of space."

The gravitational theory of Isaac Newton had dominated science for centuries. After Albert Einstein's theory of relativity around 1915, some people "rejected" Isaac Newton's theory, although it still explains many things on Earth. Before my findings, the explanation most scientists gave for gravity was based on the theory Albert Einstein wrote about a century ago.

Isaac Newton alleged that gravity is the "force that causes any two bodies to be attracted to each other, with the force believed to be proportional to the product of their masses and inversely proportional to the square of the distance between them." Putting this another way, "Newton's theory of gravitation postulated that the force between two bodies is proportional to the product of their masses and the inverse square of their separation and that the force depends on nothing else." However, a few centuries later, Albert Einstein, in his general theory of relativity, labeled gravity "not as a force (as viewed in Newton's law of universal gravitation), but as a consequence of the curvature of spacetime that he believed to be caused by the uneven distribution of mass." Einstein's theory of general relativity assumed that the "acceleration due to gravity is a purely geometric consequence of the properties of space-time in the neighborhood of attracting masses." Following Albert Einstein's theory of relativity, most classical investigations into gravity have been halted or quickly abandoned, as many believe the theory has solved the old problem of gravity's cause. Despite its imperfection, up until today, scientists have mostly used Albert Einstein's theory of relativity to try to explain celestial bodies beyond the Earth. In contrast, on the scale of the Earth, the classical theory of Isaac Newton is mostly accepted. When I learned about this discrepancy in 2013, I knew that something must be wrong with both theories. Because I was unsatisfied with the existing theories on gravity, I decided to explain gravity according to what I discovered about the mother of all turbulence, which molded the universe. Before I present my theory of gravity, I will first show what is known about the gravity of the celestial bodies.

While the gravity of the Earth can be easily measured by, for instance, throwing some objects in the air and tracking their speed over time, for other celestial bodies in the Solar System and beyond, this task is more difficult. Besides the Moon, no other celestial body has been visited by human beings in a way that would allow gravity experiments done on Earth to be tested there. Therefore, the gravities reported for most celestial bodies are based on theories forged about the formation of the universe. Because most of those theories were not based on the turbulence at the origin of the universe (which I finally decoded for the first time in history), they failed to properly estimate the true gravity of celestial bodies.

The unit of gravity is m/s^2 because it expresses the rate of change or rate of variation of speed according to time. It is like dividing speed (expressed in m/s, meaning meters per second) by time (expressed in seconds). Hence, m/s divided by second equals m/s^2. The highest gravity in the Solar System ($274\ m/s^2$) was obtained with the Sun. Among the planets in the Solar System, Jupiter (the largest) has the strongest gravity (24.79 m/s²). The gravity of the Earth is 9.798 m/s². This means the Sun's gravity is 27.96 times Earth's. The gravity of the Earth is about 6 times that of the Moon (1.62 m/s²). In general, the gravity of the satellites is less than that of their primary planets. Finally, although gravity is connected to the size of the celestial bodies, size is not the only factor that defines the gravity of the celestial bodies.

Another Book by Nathanael-Israel Israel:
RECONCILING SCIENCE AND CREATION ACCURATELY

THERE IS ONLY ONE SIMPLE, COMPELLING, SOLUTION-DIRECTED SCIENTIFIC FORMULA ACCURATE ENOUGH TO RATIONALLY EXPLAIN HOW GOD CREATED THE UNIVERSE

"Reconciling Science and Creation Accurately" is a landmark book in universe-origin writing from a rare perspective by one of the most respected minds of our time. It scientifically explores the most challenging questions of all times that believers, nonbelievers, and all freethinkers are interested in: How can we rationally demonstrate, without checking our brain at the door in the name of faith, that God created the universe? How did the universe begin and what processes did God use to create it? Are these processes still operating in the universe or not? Can believers abandon wrong theories if they think it is impossible for science to literally prove the Genesis story, or if they think that science is evil and diametrically opposed to faith, or if they compromisingly embrace scientific theories that contradict the Biblical account of creation written before the scientific era? What can believers do to help the skeptics believe in the Biblical narrative of creation?

Lucky you, Dr. Nathanael-Israel Israel successfully navigated all those questions with an accuracy that both scientists and nonscientists have been applauding across the globe. After reading *"Reconciling Science and Creation Accurately"*, you will confidently:

- Scientifically prove the Biblical account of the creation of the universe and the existence of God in a way that makes the head of those who deny God to spin faster than a DJ's turntable
- Know how to rationally talk to anti-creationists, evolutionists, Big Bang proponents, atheists, skeptics, and other freethinkers about the universe-formation and they will beg you to know more about God, the Creator, that they mistakenly rejected
- Discover very accurate, rare, and factual truths about the universe-origin that will save you time and money, and get you much closer to the better and joyful life you want to live today and forever
- Improve your health and faith by knowing that the existence of God can be scientifically justified using Science180 Cosmology and particularly Science180 Creationism

- Enter a new area of freedom and power by crushing the head of and breaking free from the suffocating expectations of all wrong theories that have highjacked secular and religious education, and that have held the Biblical account of creation captive for almost 3500 years

- Break free from the suffocating expectations of some forms of creationism that have sequestered the mind of some believers for a long time

- Uncompromisingly, intelligently, and scientifically explode the myth of those who, instead of literally taking the Biblical days of creation as 24-hours consecutive days, think that they were millions of years, or were representative of long ages, or that millions of years existed before them or were positioned between them

- Understand the accurate standard to interpret the Biblical account of creation thanks to Science180's breakthrough that transformed science and laid a foundational bedrock for the inerrancy of Scripture

Now that Genesis (the oldest manuscript in the world, written before science and most religions were born) is scientifically proven to be correct (*Science180.com*/biblical), what unstoppable, jaw-dropping paradigm shift will the discovery of the perfect alignment between science and the Bible bring into the religious, rational, and secular world today? Get this thoughtful book now to figure out what happened at the beginning, what is coming up, and why it is time to urgently rethink everything you have been told about the universe-origin so you don't eventually regret! Don't say nobody told you!

Founder of Science180 Academy, **Dr. Nathanael-Israel Israel** is acknowledged worldwide as the discoverer of the all-in-one, proven, and simple scientific formula that accurately cracked the origin of the universe, of life, and of chemicals, and that scientifically unearthed the holy grail at the intersection of science and the Biblical account of creation. Learn more at Israel120.com.

6.2. The gigantic problems hanging over the heads of the existing theories of gravity

When human beings perceive interactions between living and nonliving things, they tend to quickly view them through the lens of forces. When nations and their people interact, they usually tend to see who is stronger, more powerful, more dominant, and richer than whom. In other words, in human societies, people seem to perceive relationships between people as matters of strength, power, and influence, involving forces at play. When people see a magnet apparently "attracting" certain things (e.g., metals), they think about forces. Likewise, when people see particles and celestial bodies orbiting or moving around others, they tend to think about forces whose intensity they vehemently try to measure, and they always try to come up with some

Nathanael-Israel Israel: Member of the American Association for the Advancement of Science

numbers to express the intensity of those so-called forces, even if they do not exist. And when they "fail" (I wonder if they ever really succeeded) to model those so-called forces, they are frustrated. Yet not everything or every interaction in nature is mediated by forces. In other words, in the name of traditional mathematics and other scientific ways of analyzing data, most scientists seem to be victims of a mindset that makes them think about equations, formulas, forces, and other quantitative measurements in most of their modeling, but unfortunately, they miss the point of the problem, which is beyond math, even beyond complex math. This way of thinking worsens when people try to explain the existence of the so-called fundamental forces between particles and celestial bodies as if they are mediated by particles. For instance, when some scientists saw the associations between particles and systems of bodies on different scales, such as subatomic particles, atoms, molecules, chemical compounds, minerals, rocks, and celestial bodies, they viewed them through the lens of forces and particles mediating them. Therefore, for centuries, efforts have been made to identify the particles and forces involved in the system of particles and celestial bodies in the universe. Toward that end, several mathematical formulas have been developed to try to "summarize" (but they ended up complicating) the vast amounts of data collected on phenomena in nature and in the laboratory. The theories about the fundamental forces in nature are part of the efforts to explain nature using mathematical models involving forces and particles alleged to mediate them. In the remainder of this chapter, I will not engage in traditional or conventional mathematics that seeks to find complicated formulas to explain gravity. Instead, considering what I have amply explained about the mother of all turbulences, I will highlight what I think can help you understand gravity's perspective on turbulence with respect to the split-gathering of the precursors of bodies in the universe.

Following years of impartial analysis of the immense data collected on celestial bodies and microscopic particles in the universe, I realized that the theories of gravity advocated by Isaac Newton and Albert Einstein are wrong, although parts of them apply to certain phenomena in nature. When I considered the amount of data that I used to address the origin of the universe, I understood that nothing called "action at a distance" (as postulated by Isaac Newton) or "curvature of space" (as postulated by Albert Einstein) applies to the physics that prevailed during and after the formation of the universe. For instance, although the orbits of most celestial bodies are curved, space itself is not curved, nor does it define gravity. The curvature of space advocated by Albert Einstein is an abstract mathematical hypothesis or an attempt to geometrically explain a physical phenomenon without physical proof. Similarly, although some of the equations of the speed, radius, movement, and spacing characterizing the celestial bodies can be twisted to deduce some ideas of attraction between some bodies, the celestial bodies are not attracting one another, although the field around them can sometimes act as though they are attracting smaller things in their vicinity, but not on the astronomical scale of the celestial bodies.

Although most people, including the "top" scientists, would say that gravity has been explained and that every measurement and test on the theory of gravity crafted by Albert Einstein has been confirmed, the truth (that many experts agree with) is that gravity is still unexplained, and many problems related to it are still mysterious. Most well-informed, educated scientists know that gravity is not yet explained. For instance, although the General Theory of Relativity sought to describe the geometry of gravity, many scientists speculated that it does not capture the underlying mechanics of gravity. One of the areas where problems still arise, and which suggests that existing gravity models are wrong, is the so-called "flyby anomaly." Let me explain. Indeed, a flyby is a technique used to fly space engines by celestial bodies in order to increase their speed on their way to other celestial bodies in the Solar System. If an artificial spacecraft is sent to a very remote celestial body, such as Saturn, its speed decreases on the way. To increase its speed, such a spacecraft can be flown by a celestial body between the Earth and Saturn. The details of this technique may be hard to explain, but it involves the calculation of the gravity, speed, and size of the celestial bodies. Also called a gravitational slingshot, gravitational assist, gravity assist maneuver, or swing-by, gravity assist is a technique that uses the movement and gravity of a celestial body (e.g., a planet) to modify the path and speed (e.g., accelerate or decelerate) of a spacecraft and to save propellant, time, and other expenses. For instance, when a spacecraft approaches a planet from which it will receive a gravity assist, the planet is believed to pull the spacecraft, thereby increasing its speed. This technique was used to increase the speed of several spacecraft used to explore deep into space. For instance, such a process was used to increase the speed of the two Voyager probes (Voyager 1 and Voyager 2) by flying them by Jupiter and Saturn on their way to deep space. Since the last century, many flyby missions have been conducted, but the results have revealed numerous anomalies. Hence the term "flyby anomaly". Some of those anomalies involve artificial spaceships exceeding expected speeds. Because the gravity of the celestial bodies by which the spaceships perform the flyby is the main parameter involved in the calculation, common sense suggests that if the speed of the spaceships increases beyond the expected value, the gravity of the celestial bodies must have been wrongly calculated, implying that the gravity calculation used is wrong. Putting it another way, according to the "flyby anomaly", many spacecrafts have experienced greater acceleration than expected during gravity assist maneuvers, suggesting that gravity may have been underestimated. Yet, despite this kind of proof or alert, people still wrongly think the gravity theories are still correct! Therefore, nothing has been done to correct most theories about the origin of the universe, which were mistakenly anchored to or strongly connected with theories of gravity.

6.3. Thinking that Albert Einstein and Isaac Newton explained what gravity really means? What if they were all wrong? Read this to find out!

If I had not decoded the mysteries surrounding the mother of all turbulences, I could not have decrypted how gravity was established and what it means. In my book *"Turbulent Origin of the Universe,"* I provided a lot of details, but for the sake of space in this book, I will just pinpoint a few things. Indeed, gravity relates to the turbulent vortical movement (think of it like a swirling, spiraling, or winding up) of the structures (on a small and large scale) in the turbulent fluid flow during the formation of the celestial bodies. Gravity is a consequence of the process that split-gathered the precursors of celestial bodies into their daughter bodies. Without a mechanism, process, or force to combine the precursors into unified bodies, most daughter bodies could not have formed after their split from their mothers. The gathering together of the matter in the precursors of bodies as they started splitting from their mother was not initially controlled by gravity. It was later that gravity itself was established. Put another way, before what is known today as gravity was established, forces and interactions were already acting on the precursors of bodies, helping gather their matter into unified bodies and systems of bodies. Gravity was like the last state of expression of some of the forces that acted to collect dispersed matter in the precursors of bodies. If the processes that formed the universe could be reversed and the celestial bodies unwound, the universe could be brought back to a state near that of the initial matter. For the bodies in the universe to be formed, a process was needed to split and pull together their precursors into different "compartments."

The way fluid layers and the vortical structures formed in them were moved, collected some precursors into bigger bodies and others into smaller ones, according to the split-gathering followed by laws I extensively explained in my books on the origin of the universe. For some bodies to get their size, fluids surrounding their precursors were accumulated as they were moving until they reached a position where they could no longer be sheared any farther by the flow. Consequently, fluid bodies were "stationed" at positions that agreed with their orbital speed, rotational speed, and other characteristics. In other words, the force that sheared the fluids was not able to move their bodies farther away beyond some limits defined by their energy, location, and other factors. Thus, each body was positioned at a location matching its size, movement, constitution, and the split-gathering history of its precursor. Just as the wind can easily move a leaf but not a heavy house made of concrete, so also, when a big body was formed in the way of the fluid flow, the ability of the flow to displace or position it depended on many factors, including the energy, the stress in the flow, and the energy of the body itself. But if the body contains more energy, it may not be moved by the flow beyond a certain distance. Hence, the largest bodies are not usually very far from their mother precursors. Similarly, the energy that could have caused the shear flow could have had a power that pulled or pushed things in its way or into its surroundings. This is like how a tornado knocks out things in its way, both small and large. The viscosity of the precursors could have also affected their stretching. As all of this was happening, bodies were being rotated, and gravity was progressively formed according to the rotation and size of the bodies' precursors. As of today, artificial gravity can be "created" using the rotation of bodies, therefore suggesting that natural gravity itself could as well have been the consequence of the rotation of the precursors of the celestial bodies.

Under the influence of turbulence, vortical structures present in fluid layers were progressively squeezed, amassed, and wound up into bigger bodies as fluids were pulled and pushed while movements were being born. Some precursors of bodies started rotating more than others. Gravity was progressively established as the fluids of the precursors were amassed under the influence of their movement (e.g., rotation or spiraling). The forces involved in the flow of the precursor fluids compressed the latter according to the intensity of their rotation and size, and consequently affected the gravity of their daughter bodies. In other words, gravity and its precursor prevented the precursors of celestial bodies from dispersing but formed unified bodies. Another way to explain this complex process is that, as the precursors of the celestial bodies were being split and moved, a squeezing force began to compress and shape them as internal processes gathered them into clusters of rocks, minerals, atoms, subatomic particles, and other smaller constituents. Because gravity itself depends on the radius and rotational speed of the celestial bodies, its intensity as of today could not have been established before the bodies got their current shape. That is to say that, as the precursors of the bodies were taking shape, size, and speed, a process of gathering them and establishing gravity began. Just as all bodies today have had their own precursors, so also all forces and interactions in nature have had their precursors. Similarly, gravity had its precursor, which was embedded in how turbulence allowed fluids to be gathered together on astronomical scales. By the setting of the size, movement, and orbit of the celestial bodies, gravity was established.

Without a mechanism to match gravity and the other systems of gathering together of the precursors with the size, movement, and composition of the celestial bodies, some bodies could have been highly compressed, while others could have been scattered or dispersed. As they were moving inside the fluid layers of the precursors of bodies, the vortical structures could have pulled some bodies just as rotating vortices can pull matter toward themselves. By the time the celestial bodies were formed, a seemingly "attractive" field or zone was created around them, and most things that are in that zone can be pulled toward the celestial bodies. Even at the atomic and subatomic level, there is a field around particles. It is this field that allowed particles to exist or maintain their cohesion, and without it, matter would have had a hard time.

Gravity can also be affected by the air, gases, and waves coming from the poles due to the rotation of the celestial bodies. In other words, the rotation of the celestial bodies and even of any other body can cause the formation of a field around it, and the intensity of that field can depend on the size, rotational speed, and other factors related to the bodies. For instance, when fluids in a bowl are rotated, a force can be born that causes the formation of a conical structure at the surface of the fluids and attracts some air from the outside toward the inside of the fluids. A similar effect is found with hurricanes and many other moving things. Gravity is higher at the poles because the air is pulled toward the poles directly, whereas it is lower at the equator because the equator is farther from the poles. The force of the air descending at the poles is accompanied by winds, which explains why the poles are colder than the equator, although some of the poles are in contact with the Sun more than the equator.

Gravity is like the leftover of the force that defines the existence and organization of the celestial bodies. To explain this concept, I drew a comparison from living things. Indeed, every living organism and nonliving thing seeks to protect its domain. Human beings, wild animals, and planets do the same. On the national level, countries and states try to protect their borders and neighborhoods. In each nation, borders exist to delimitate states, counties, cities, or towns. On the individual level, human beings like to hide themselves behind fences and protect themselves with clothes to withstand adverse weather. Without these strategies, human beings can be destroyed by perilous environmental conditions. From the herbivores to the carnivores, wild animals have their ways of protecting themselves according to their ecological niche. Although the carnivores try to dominate the herbivores, while bigger animals try to dominate smaller ones, each wild species manages to have a territory to enjoy. When strangers enter the property of someone else, they can be attacked and charged with trespassing, or even worse, killed. Some immigrants are sometimes caught, detained, and deported from their new country of residence if they do not have their immigration papers properly set. When a stranger enters a new environment or niche of a different species, the latter can oppose, attack, destroy, or try to dominate the former. This is not true just for human beings but also for wild animals and plants. Some wild animals can kill others just to defend their territory. Some of these behaviors hide a fundamental code or law of defense not only of ecological niches but also of existence, sustaining life amid environmental challenges and maintaining underlying global laws governing the formation of the universe. Because they cannot move from one place to another on their own, plants adapt to their environment. They also know how to defend their ecological niche and even compete with other plants in their environment. Plants of the same species seem to enjoy living together more than living with plants of a different species.

As far as nonliving things are concerned, matter is governed by a law that allows it to defend its environment. Although they may not be as conscious as some familiar living things, atoms seem to "know" how to maintain their integrity and not "voluntarily" engage themselves in reactions or interactions that will destroy them. For instance, not all atoms associate with all other atoms. And when a "stranger" enters their environment, "frustrations" manifest as reactions, including radioactive ones, which can harm nearby living organisms. Likewise, on the astronomical scale, celestial bodies (e.g., satellites, planets, and stars) know how to defend their environment or propriety. When a stranger enters their environment, celestial bodies can attract them and pull them toward their surface using gravity. In other words, gravity is one of the mechanisms the celestial bodies "use" to maintain and "defend" their environment or property in the universe. The imparted energy, motion, and other characteristics to things in nature were not meant to be lost easily, nor was the force to reverse these laws cheap!

As I carefully studied bodies in the universe, I realized that gravity is not the force that explains the revolution of celestial bodies, but rather a force that arises from how the precursors of these bodies were assembled, enabling them to maintain their existence and cohesion. It is a major error that some scientists have made by thinking that planets attract their satellites or that the Sun attracts the planets, and vice versa. The misunderstanding of gravity contributed to the scientific drifting in, for instance, astrophysics and cosmology, which seems to have become a religion in which things are claimed, rejected, and accepted without proof using some confusing statistical and other mathematical tools in the name of physical things we cannot see. On the scale of subatomic particles and atoms, the process that formed them and defined their limits also locked them into compartments in such a way that some theorists can mistakenly think that those particles are attracting one another. In other words, just as some theories on gravity make people believe in an "action at a distance" (e.g., hypothetical attraction between a primary body and its secondary bodies), some people mistakenly think that nucleons and electrons (particles found inside and outside the nucleus of an atom, respectively) are attracting one another. In my book on the formation of the chemical particles, I provided more details. Moreover, contrary to what most theories claim, gravity will never be sufficient to explain the motion and functioning of celestial bodies and their clusters. Anything that can reverse the direction of the rotation and revolution of the celestial bodies, meaning causing celestial bodies to rotate and orbit in a direction and sense contrary to what they are as of today, could also reverse gravity. The systems or processes that progressively established gravity were connected to those that established rotation, revolution, and many other properties of celestial bodies and particles in the universe. But because human beings and the level of their technology are not and will not be able to reverse the course of the movement of the celestial bodies, they can neither change the gravity of the celestial bodies nor change the design, functioning, and the course of the history, story, and destiny of the universe. I talked about that destiny in other books.

Another Book by Nathanael-Israel Israel:
HOW BABY UNIVERSE WAS BORN

If you don't believe in God or you hate God, or you don't think there is anything or anyone called God, but you want your children to understand how the universe was formed from a scientifically-proven perspective that considers the facts, then this book is for your children.

Dr. Nathanael-Israel Israel is the founder of Science180, the American organization that helps people enter the realm of true knowledge about the universe formation. In other words, he is known as the first human being to ever use modern science to give people the state-of-the-art decoding experience of the origin of the universe and of life.

Another Book by Nathanael-Israel Israel:
HOW GOD CREATED BABY UNIVERSE

THE FIRST AND ONLY BOOK THAT ACCURATELY EXPLAINS EVERYTHING ABOUT THE FORMATION OF THE UNIVERSE AND LIFE IN A WONDERFUL LANGUAGE THAT ALL CHILDREN AGES 7-12 CAN EASILY, FULLY UNDERSTAND & ENJOY!

As the only universe-origin book that your whole family will like and enjoy together, *"How God Created Baby Universe*" will set children on the path of success by accurately showing them early in life the formation of the universe, and how to detect errors in theories or stories that would misguide them as they grow up. Therefore, you need to add this great, efficient, trustworthy, and cost-effective book to the strategic journey of children toward their best tomorrow. With *"How God Created Baby Universe"*, you will:

- Have a peace of mind that children will get accurate, fit, and easy to understand universe-origin information that will produce real results in their life

- Become the leader that captures the heart of children craving for the original explanation of the formation of the universe so you can clear their way for freedom, power, technology, innovation, and breakthroughs of the future (learn more at Science180.com/children)

- Protect yourself and loved ones from wrong theories in the literature and the media by keeping children secured and empowered with the true knowledge of how the universe began

- Explain complicated secrets to children about how to locate mistakes in origin-related theories so you can save time, money, and other resources to improve their lives

- Ultimately boost children's confidence in detecting, confronting, and avoiding wrong theories by knowing the facts and real processes involved in the formation of the universe

- Help children to easily sort out their origin-related questions using strategies that get them to tap into deep secrets that even highly educated people ignore

- Clearly explain to children how to mathematically know without a doubt whether God created the universe as the Bible says or billions of years evolution processes formed it

Accurately explaining the complex formation of the universe and of life to children can be very hard in our modern world, but by getting *"How God Created Baby Universe"*, you will know the proven formula to help children to easily understand their huge universe-origin and life-origin questions with confidence, humor, and joy.

They will surely laugh aloud while reading this book and thank you for it! It is time to buy this pragmatic book to help the children in your life today.

Member of the American Association for the Advancement of Science, American Chemical Society, and the American Society for Microbiology, **Dr. Nathanael-Israel Israel is** a Beninese-American scientist and international consultant, who shows the world how to scientifically decode the formation of the universe, of life, and who is known as the creator of the Chemicals Turbulent Origin Formula™, the inventor of the Life Turbulent Origin Formula™, and the discoverer of the Universe Creation Formula™. Learn more at Israel120.com.

6.4. Are we giving the "fundamental forces" in nature a bad interpretation?

In the previous segments, I placed the accent on gravity, but here I will quickly address other so-called fundamental forces. Although countless studies have been performed on atoms, nobody has ever physically seen a single atom or electron yet. But when countless them are combined, human beings can sense them. For instance, although it is well known that water (chemically annotated H_2O) that we drink every day consists of 2 atoms of hydrogen (annotated H) and one atom of oxygen (annotated O), no one has ever seen a single hydrogen atom or a single oxygen atom yet, even by using the most advanced microscope. All existing theories of atoms are based on effects felt on them and the manifestation of their properties or damage caused to them. This also implies that most theories about the nature of chemical particles are not 100% based on reality but rather on assumptions that are largely invisible to the naked eye.

If scientists cannot properly explain gravity, which can be easily measured with tools that can be seen, how can we trust their so-called measurements and theories done on smaller particles like atoms and subatomic particles that nobody has ever seen? I demonstrated in *"Turbulent Origin of the Universe"* that as the turbulent prima materia was pushed into motion, destabilized, and entered turbulence, different fields could have been birthed and formed across the compartments of matter being split-gathered. These fields could have been the precursors of some interactions that were twisted by theorists to invent the so-called fundamental forces. Although it can be okay to talk about forces within atoms, I am not convinced that they are mediated by particles, but rather by the processes that determined how matter and its clusters were formed. This way of thinking aligns with the explanation of gravity, which does not mean that the constituents of the thing in my hands are attracting one another. The interactions between human beings through love and sometimes hate can be strong and, for some, seem mediated by a force of love or hatred, yet they are not mediated by a particle, even if the settings of the human heart may play a role.

Similarly, by the end of the universe's formation, natural "forces" may seem to be present between bodies, but they were not the underlying cause of the universe's formation. Some of the "forces" that acted on the precursors of the bodies and still act on their daughter bodies today have been products of their movement, including rotation and revolution. For instance, as bodies rotate, they can attract some air toward their poles. Consequently, the rotation of the celestial bodies can be responsible for the influx of cold air toward their poles. The forces accompanying the incoming air at the poles can also explain why gravity is higher at the poles, why the poles are flatter than the equator, and why the equator is oblate.

Unlike what some people think, the so-called fundamental forces were not instantaneously forged in one step. However, as the structures and characteristics of the bodies' precursors in the early universe were changing, so too did the intensity and nature of the interactions between them, some of which would later be called fundamental forces. The characteristics of these forces and the things that mediate them evolved until a kind of equilibrium was reached, and most of the constituents of matter (small and big) were locked into systems whose dynamism is much smaller than what prevailed at the beginning of the world.

If the interactions between particles and celestial bodies can be properly inventoried, more than four fundamental forces will be found in nature. As the gravity of a planet in the Solar System is smaller than the gravity of the Sun, so also the gravity of the Sun itself may be smaller than the gravity of the core of the Milky Way Galaxy. Similarly, at the scale of galaxy clusters, interactions other than gravity can explain the organization of their structures. In other words, while on an astronomical scale, gravity may be used to explain the field around a single body, on the scale of systems of bodies such as stars, galaxies, and their clusters, gravity can be inappropriate. To be accurate, a new concept different from gravity must be used to explain the mechanism holding together the clusters of bodies, such as:

- A stellar system,
- A galaxy,
- A cluster of galaxies,
- A cluster of clusters of galaxies, etc., and
- All the clusters of galaxies in the universe.

I believe that the way I explained gravity will usher in a new era in science and across all scientific domains where gravity is used. In my book *"Reconciling Science and Creation Accurately*," I also explained why and how two types of gravity exist: physical gravity and spiritual gravity. But in this chapter, I mostly focused on the physical gravity, although at times, I also apostrophized the spiritual ones that some people would not want to pay attention to if they were presented in a religious way!

In *"Turbulent Origin of the Universe"* and in *"Turbulent Origin of Chemical Particles*," I extensively reviewed the so-called fundamental forces and the so-called Standard

Model theory of particles, which is alleged to connect most of the known particles together.

6.5. Take-home message

In this chapter, I provided a new perspective on gravity. I explained that gravity was born as the turbulent processes that formed the celestial bodies were gathering together and moving the fluids in the precursors. For instance, I showed that, just as a fluid filament can be curled by a current to form vortices, which can rotate and attract other fluids, so also as the precursors of celestial bodies were being formed, their fluid layers swirled at one point and conferred a rotation to their daughter celestial bodies, with an ability to attract neighboring matter. I showed that the so-called fundamental forces are not the only (fundamental) forces in the universe. Some of them are not even worth calling forces, and most of them do not even really explain reality as seen in nature. Based on my investigation, I perceived that gravity is not responsible for or does not control the orbit or trajectory of the celestial bodies. The precursors of the celestial bodies had already been launched and were on their way to their orbits before gravity was fully established. Gravity is neither the process that causes secondary bodies to orbit their primary bodies, nor is it an attractive force between all matter. In the Solar System, for instance, gravity is NOT responsible for the revolution of either the planets or the asteroids around the Sun, nor for the revolution of the satellites around their primary planet. Similarly, gravity is not responsible for the revolution of the Sun around the center of the Milky Way Galaxy. Gravity is not responsible for the large-scale structures in the Universe, but the way the bulk of the turbulent prima materia (the precursor of all matters in the universe) was split-gathered is. Gravity is responsible neither for the constitution of celestial bodies nor for their mass. The mass of the bodies in the universe was defined by the amount of the turbulent prima materia that was amassed and squeezed to birth them. More details about gravity can be found in my books on the origin of the universe.

Another Book by Nathanael-Israel Israel:
TURBULENT ORIGIN OF LIFE

THE ONLY ACCURATE FORMULA TO SCIENTIFICALLY EXPLAIN THE FORMATION OF ALL FORMS OF LIFE QUICKLY

Every human being will benefit from understanding the real origin of life. But the problem is that most efforts to explain the origin of life are complex, inaccurate, confusing, partisan, complicated, therefore, creating serious challenges to those who are eager to scientifically decrypt where all forms of life came from. Most people want an accurate, simple, straightforward, nonpartisan life-origin book that is free from jargons and difficult concepts only known by the experts. This elegant scientific book breaks down the technicality of the origin of life in a language that even the nonscientists can easily comprehend. It is a trustworthy book that will help you to quickly, cheaply, easily, and efficiently navigate everything you need to know to finally decode and solve the puzzling problems about the origin of life, while also giving you a crash course on the universe-origin.

Unlike any book you have ever read on the origin of life, this historic masterpiece (that distills complex scientific data down to simple explanations that make sense) is the starting point of any smart person wanting to rationally understand the formation of all living things. By the time you finish reading *"Turbulent Origin of Life"*, you will discover:

- Why in spite of the massive amount of scientific data collected on living things, scientists have misunderstood the formation of life until now, and then uncover in a simple language the one thing that was needed to accurately crack the code of life but that scientists have missed and that has been causing them headaches, overwhelm, and burnout

- Step-by-step pathway to decode the origin of life and get the power, freedom, and boldness to take advantage of the opportunities that accurate understanding of the origin of life creates (*Science180.com*/life)

- The high connection between the code of the universe formation and the process by which life on Earth was formed so you can become a fulfilled thought leader in your field of expertise

- Tools to stand as a lighting bolt that electrifies those who are still struggling to understand the formation of all forms of life in the universe

- Strategies to push the boundaries of human abilities to properly understand what is perceived as un-understandable, mysterious, supernatural, unimaginable, impossible, and unthinkable that hold people back
- Scientific approach to holistically detect, correct, and remove all misinformation, ambiguity, and misleading claims and theories surrounding the origin of life

Whether you are a scientist or a layperson, a believer or a skeptic, you cannot afford to ignore the greater, better, faster, simpler, cheaper, easier, and accurate formula unlocked in this important book that successfully decoded the origin of life. Get *"Turbulence Origin of Life"* today and change lives! Don't wait!

Dr. Nathanael-Israel Israel is the Father of Science180 Cosmology and the Founder of Science180 Academy. He is fortunate to be known as the source of unconventional wisdom and knowledge that help people accurately crack the code of the formation of the universe, of life, and of chemicals. Get some resources by visiting his personal website at Israel120.com.

'Science180 Academy' Success Strategy
SCIENCE180 PUBLISHING: AUTHORS WANTED

Science180 Publishing, the American publishing company that published the groundbreaking discovery about the origin of the universe, of life, and of chemicals spearheaded by Dr. Nathanael-Israel Israel, really wants to publish your book(s) regardless of your field of expertise. This is a unique opportunity for:

- established authors
- people aspiring to become authors
- people who have written a book or are wanting to write one and need help with anything regarding publishing
- people who are not well known, inexperienced
- people whose books are viewed as nonconformist, controversial, or unconventional
- people who do not have enough resources or knowledge to navigate the publishing process
- people who are struggling to find an affordable, experienced, and high-quality publisher

Although Science180 Publishing is based in the USA, it can publish your books within your budget regardless of your geographical location. Science180 Publishing is highly interested in your document and possibly helping you publish it. Please visit Science180Publishing.com to explore how we may assist you. No matter the content of your book, as far as it is original, not promoting anything illegal, not duplicating anyone else's idea, Science180 Publishing can help you publish it in the USA. Please contact us asap and see how we can help.

To start your journey of publishing your book with Science180 Publishing, please visit Science180Publishing.com today.

'Science180 Academy' Success Strategy:
SCIENCE180 INTERVIEW REPORT (AKA SCIENCE180 INTERNET-TV-RADIO INTERVIEW REPORT)

Science180 Interview Report is the newsletter to read for guests and unconventional show ideas at the intersection of science and faith. Indeed, many hot questions are still unanswered on the road leading to the correct understanding of the origin of the universe, of life, and of chemicals. But most people don't know where to find the accurate answers to those challenging questions. What if with one simple call you can accurately answer all of those questions. You need to get in touch with or interview Dr. Nathanael-Israel Israel on your show, radio, tv, podcast, and even website, or invite him for a live presentation at your organization if your audience can benefit from any of the following show, talk, speaking, or interview ideas:

- Are most Christians denying the God they want the nonbelievers to accept?
- Can anyone really be scientifically 100% sure and prove that God created the universe?
- Can mathematics and science collide to accurately explain the creation of the universe, of life, and of chemical particles?
- Can most Christian leaders refuse to take a stand on 6 literal days of creation and expect atheists and freethinkers not to argue that God is simply unnecessary?
- Can we explain the formation of the universe through natural processes without evoking evolution and Big Bang?
- Can we explain the origin of the universe, life, and chemicals through natural processes without evoking evolution and Big Bang?
- Can we mathematically prove that the formation of the Earth was completed on the 3rd day of creation like the Bible says?
- Can we scientifically demonstrate without a doubt that the Moon and the Sun were really formed on the 4th day of creation like the Bible says?
- Can you be really free from doubt about the universe-origin and God's existence?
- Did the Quran and any other religious book make any gigantic error about the universe creation that any scientific formula proves the Bible got right?
- Does the Bible scientifically teach anything about the universe-origin that most people including Christians ignore?
- Does the Bible scientifically teach anything about the universe-origin that most Christians ignore?

- How can people approach difficult topics such as challenging the accuracy of the Biblical account of creation, the Big Bang theory, or any all other theory on the origin of the universe?
- How can people benefit from understanding the mother of all turbulences?
- How people, including some fervent Christians, come to believe lies about creation and what they can do to change them so atheists can enjoy God.
- How to raise rational children in our modern world?
- How to scientifically prove without talking about the Bible that God created the universe
- Is faith better than science?
- Is it a waste of time to attempt to prove the Bible is true by means of science or historical investigations?
- Is it a waste of time to attempt to prove the Biblical creation using science or historical investigation?
- Is science making you doubt your faith or is the Biblical account of creation making you doubt science?
- Is the war between mathematics and physics drifting science in the wrong direction?
- Must Christians apologize to atheists, rationalists, and all freethinkers for the proofs creationist scholars and preachers have used to argue creation?
- What are the 3 world-shaking truths about the separation of science and faith nobody ever told you?
- What are the 4 surprising essential skills for smart people to crack the universe-origin?
- What can scientists and nonscientists do to understand the origin of the universe in a polarized world, whether it's at church or at public schools?
- Why arguments against secular science are NOT arguments for creation
- Why Christians are abandoning wrong creationist theories that compromise with Darwinism and Big Bang?
- Why do secular rationalists and freethinkers think that Christians are irrational?
- Why the secularist world doesn't care much if Christians and their leaders believe in Evolutionism, but they actually care much if they don't believe in the billions of years process?

I know you may be tempted to answer these questions by yourselves, but avoid landing yourself on wrong paths that caused some people to lose contact with reality, it is better to get the accurate answer from the know-how expert, Dr. Nathanael-Israel Israel, the author of many books on the origin of the universe, of life, and of chemicals, and the standout expert who accurately decoded the scientific formula that forces science to bow to the truth. If you would like to register to Science180 Interview Report so we can periodically send you show ideas and opportunities related to the origin of the universe, of life, and of chemicals particles, please visit Science180Interviews.com for more details. To get Nathanael-Israel Israel to answer any of these questions for you, please contact him at Israel120.com.

CHAPTER 7

CAN ANYONE REALLY BE SCIENTIFICALLY 100% SURE THAT GOD CREATED THE UNIVERSE? PAY ATTENTION TO THE SCIENCE180 MODEL OF COSMOLOGY FOR THE GENERAL PUBLIC!

Can any scientific theory successfully challenge the celebrated "general theory of relativity" (also known as "Einstein's theory of gravity") and the nearly abandoned classical physics of Isaac Newton?

You are worthy to understand the origin of the universe, and if you have read this book from the beginning until this point, your perspective of the beginning must have drastically improved. As a summary of the most comprehensive theory ever elaborated for the benefit of anyone who is seeking to know the true origin of the universe, this book advances human understanding of the universe by:

- providing a paradigm shift of the underlying factors responsible for the trends of several variables collected on celestial bodies,
- helping to reconcile the longstanding, apparent contradictions between the scientific evidence and the Biblical creation narrative of celestial bodies (e.g., Earth, Moon, and Sun), and
- laying the foundation for future research to identify new hypotheses that can reevaluate many existing models in physics and in other fields involving turbulence and fluid breakup.

Although the origin of the universe has long preoccupied humankind and, throughout history, efforts have been made to decode nature, many questions remain unanswered. This book answers many of them. In this manuscript, which focuses on the formation of celestial bodies (e.g., galaxies, stars, planets, asteroids, and satellites), I showed that, at the beginning of the universe, an original particle (which I termed the turbulent prima materia) appeared, and, after going through complex turbulent

Science180: Easy-to-Understand Universe-Origin, Chemicals-Origin, and Life-Origin Theory

processes, it birthed the precursors of all kinds of bodies or matter in the world. Indeed, soon after its mysterious appearance, the bulk of the turbulent prima materia was broken apart, and the resulting clusters were scattered in a gigantic explosion. I coined the term "split-gathering" to describe how the precursors of bodies were split and then gathered as the universe underwent its genesis. As they started moving, the chunks of matter underwent a cascade of breakups, during which the daughter bodies were also gathered. Some precursors of bodies were split-gathered into daughter bodies, which, in their turn, went through similar processes, and so on until the breakup stopped at the scales of the smallest particles. During the cascade of split-gathering, precursors of galaxies, including the precursor of the Milky Way Galaxy, were formed. As the precursor of the Milky Way Galaxy was being shaped, the precursors of its stellar systems were formed. The precursors of the stellar systems split-gathered into the precursors of their stars, planetary systems, and asteroid systems. In its turn, the precursor of the Solar System split, giving rise to the precursor of the Sun and the bodies orbiting it. In fact, during the split-gathering of the precursor of the Solar System, the fluids of the precursors of the bodies orbiting the Sun escaped the precursor of the Sun at about 617.6 km/s (i.e., the escape velocity of the Sun). These fluids were organized into layers, with the top layers moving faster than the layers beneath them. As they moved, the fluid layers separated and were gathered into celestial bodies. After traveling about 57,910,000 km (semi major axis of Mercury, the most renowned innermost body in the Solar System as of 2022) at about 617.6 km/s, meaning traveling for 26.05 hours (57,910,000 km / 617.6 km/s) after the beginning of the formation of the Solar System, the fluid layers of the precursor of the bodies orbiting the Sun started splitting into precursors of celestial bodies:

$$57,910,000 \text{ km} / 617.6 \text{ km/s} = 26.05 \text{ hours}$$

These 26.05 hours are what I termed the escape time of the precursor of the Sun. After its formation, the precursor of the Sun began to take shape. Due to the precursors of the bodies orbiting the Sun starting to be split-gathered into the precursors of various celestial bodies in the Solar System. Before addressing how the Sun formed, I will first discuss the formation of the Earth and the Moon.

Indeed, after the fluid layers of the precursor of the bodies orbiting the Sun split from the fluid layers of the precursor of the Sun, the first fluid layers to split were those of the precursor of Mercury. Then, the fluid layers of other celestial bodies in the Solar System followed according to their position. As the fluid layers of the precursor of one celestial body separated from the stack, the remainder of the fluid layers of the precursors of the bodies orbiting the Sun continued their journey away from the precursor of the Sun until the layers of the precursor of the next celestial body split, and so on and so forth.

The fluid layers of the precursor of the Earth-Moon system were embedded within the fluid layers of the precursor of the bodies orbiting the Sun, and they had to wait for all fluid layers above them to split or be removed before they could, in turn, separate from the rest. After the remainder of the fluid layers of the precursor of the bodies orbiting the Sun traveled about 149,600,000 km (semi major of the Earth) after escaping the precursor of the Sun at about 617.6 km/s, the precursor of the Earth-Moon system split from the fluid layers beneath it (by this time, all the fluid layers above the precursor of the Earth-Moon system split). In other words, the time elapsed since the origin before the precursor to the Earth-Moon could be formed was about the semi-major axis of the Earth (149,600,000 km) divided by the escape velocity of the Sun (617.6 km/s), meaning 67.286 hours (i.e., 2.804 days):

$$149,600,000 \text{ km} / 617.6 \text{ km/s} = 67.286 \text{ hours} = 2.804 \text{ days}$$

That is what I called the semi-major axis timescale. Within a few minutes of its formation, the precursor of the Earth-Moon system split into the precursors of the Earth and the Moon. In fact, during the split-gathering of the precursor of the Earth-Moon system, the precursor of the Moon escaped the precursor of the Earth at about 11.19 km/s, which is the Earth's escape velocity. As the fluid layers of the precursor of the Moon started their journey toward the place where they would be collected, the fluid layers of the precursor of the Earth moved at about the mean orbital speed of the Earth (i.e., 29.78 km/s) and were collected into a spherical planet called Earth (equatorial radius = 6378.137 km). Because the length of the fluid layers of the precursor of the Earth at the time of this swirling was about the circumference of the Earth, and because the swirling occurred at about the orbital speed of the Earth, the duration of the swirling of the fluid layers of the precursor of the Earth was about the circumference of the Earth (2 x 3.14 x 6378.137 km = 40054.7 km) divided by the orbital speed of the Earth (29.78 km/s), meaning 22.42 minutes.

$$2 \text{ x } 3.14 \text{ x } 6378.137 \text{ km} / 29.78 \text{ km/s} = 40054.7 \text{ km} / 29.78 \text{ km/s} = 22.42 \text{ minutes}$$

These 22.42 minutes are what I call the Earth's circumference timescale. Adding the semi-major axis timescale of the Earth (67.286 hours) and the circumference timescale of the Earth (22.42 minutes), I showed that the Earth was formed 67.66 hours (i.e., 2.82 days) after the beginning of the Solar System.

Birthdate of the Earth = 67.286 hours + 22.42 minutes = 67.66 hours = 2.82 days = on the 3rd day

In other words, the scientific evidence proved that the formation of the Earth was completed on the 3rd day after the beginning.

While the precursor of the Earth was being gathered into the Earth, the precursor of the Moon (which escaped it at about 11.19 km/s, escape velocity of the Earth), traveled about 384,400 km (semi-major axis of the Moon) before reaching a point when its final gathering together into the Moon could occur. Dividing the length of the journey of the precursor of the Moon after escaping the precursor of the Earth (i.e. semi major axis of the Moon) by the escape velocity of the Earth (11.19 km/s), I showed that in 9.54 hours (384,400 km / 11.186 km/s) after escaping the precursor of the Earth, the precursor of the Moon was set to swirl to form the Moon (radius = 1738.1 km):

$$384,400 \text{ km} / 11.186 \text{ km/s} = 9.54 \text{ hours}$$

These 9.54 hours are what I call the semi-major axis timescale of the Moon. At this point in the genesis of the Moon, the fluid layers of the precursor of the Moon, whose length was about the circumference of the Moon (2 x 3.14 x 1738.1 km = 10915.27 km) moved at about the orbital speed of the Moon (1.02316 km/s) to wrap around to form the Moon. The duration of this swirling can be calculated by dividing the circumference of the Moon (10915.27 km) by the orbital speed (1.02316 km/s) of the Moon:

$$10915.27 \text{ km} / 1.02316 \text{ km/s} = 2.96 \text{ hours}$$

In other words, after reaching the semi-major axis of the Moon, the fluid layers of the precursor of the Moon were gathered together within 2.96 hours into the Moon. These 2.96 hours are the circumference timescale of the Moon. Because it took 9.54 hours for the fluid layers of the precursor of the Moon to travel from the precursor of the Earth to the orbit of the Moon, and then another 2.96 hours for the fluid layers of the Moon to swirl and form the Moon, I proved that, after escaping the precursor of the Earth, the precursor of the Moon spent about 12.5 hours (9.54 hours + 2.96 hours) shaping itself before forming the Moon. In other words, with respect to the precursor of the Earth, the duration of time that elapsed before the Moon was formed was the sum of the semi-major axis timescale of the Moon (9.54 hours) and the circumference timescale of the Moon (2.96 hours). But considering that, after the beginning of the formation of the Solar System, about 67.286 hours had passed before the precursor of the Earth-Moon system could be formed, the duration of time that elapsed since the beginning before the Moon was formed was:

$$67.286 \text{ hours} + 12.5 \text{ hours} = 79.786 \text{ hours} = 3.324 \text{ days}$$

In other words, after the precursor of the Solar System began split-gathering, about 67.286 hours passed before the precursor of the Earth-Moon system formed and split from the fluid layers beneath it. Then, from that time, about 12.5 hours passed before the precursor of the Moon, which had escaped the precursor of the Earth, could travel to and position itself in about the Moon's orbit, then swirl to collect its fluid layers into the Moon. Hence, the duration of the formation of the Moon is equal to the sum of (1) the semi-major axis timescale of the Earth, (2) the semi-major axis timescale of the Moon, and (3) the circumference timescale of the Moon:

Birthdate of the Moon = 67.286 hours + 9.54 hours + 2.96 hours = 79.786 hours = 3.324 days = on the 4th day since the beginning

As the Earth and the Moon were forming, the fluid layers of the Sun's precursor were being molded into the Sun. Indeed, after the beginning of the formation of the Solar System, about 26.05 hours had passed before the precursor of the Sun was formed: that is, the escape time of the precursor of the Sun. By that time, the precursor of the Sun was formed, and the length of its fluid layers was about the circumference of the Sun. Considering the radius of the Sun (696,000 km), its circumference is 2 x 3.14 x 696,000 km = 4,370,880 km. The fluid layers of the precursor of the Sun swirled at about the orbital speed of the Sun (19.4 km/s), meaning that about 62.58 hours (i.e., 4,370,880 km / 19.4 km/s) after the precursor of the Sun was formed, its fluid layers were collected into the Sun:

2 x 3.14 x 696,000 km / 19.4 km/s = 4,370,880 km / 19.4 km/s = 62.58 hours

These 62.58 hours are what I termed the circumference timescale of the Sun. Because about 26.05 hours had passed before the precursor of the Sun was formed, and then about 62.58 hours had passed before the fluid layers of the precursor of the Sun could form the Sun, the total amount of time that passed since the beginning before the Sun was formed was 26.05 hours + 62.58 hours = 88.63 hours = 3.693 days.

Birthdate of the Sun = 26.05 hours + 62.58 hours = 88.63 hours = 3.693 days = on the 4th day since the beginning

In other words, the Sun was formed on the 4th day after the beginning of the formation of the Solar System.

To sum it up, the scientific evidence I presented in this book showed that the Earth was formed 2.819 days (that is, 2 days, 19 hours, 39 minutes, and 33 seconds) after the beginning. The Moon was formed 3.325 days (that is, 3 days and 7 hours and 47 minutes and 41 seconds) after the beginning, and the Sun 3.693 days (that is, 3 days and 16 hours and 37 minutes and 49 seconds) after the beginning of the formation of the Solar System, which I proved is about the beginning of the formation of the universe.

These timescales perfectly match the Biblical creation story, according to which the Earth was formed on the 3rd day, and the Moon and the Sun on the 4th day of creation (the Bible's Book of Genesis 1:1-19):

FROM SCIENCE TO BIBLE'S CONCLUSIONS

Genesis 1:1 "In the beginning, God created the heaven and the earth. 2 And the earth was without form, and void; and darkness was upon the face of the deep. And the Spirit of God moved upon the face of the waters. 3 And God said, Let there be light: and there was light. 4 And God saw the light, that it was good: and God divided the light from the darkness. 5 And God called the light Day, and the darkness he called Night. And the evening and the morning were the first day. 6 And God said, Let there be a firmament in the midst of the waters, and let it divide the waters from the waters. 7 And God made the firmament and divided the waters which were under the firmament from the waters which were above the firmament: and it was so. 8 And God called the firmament Heaven. And the evening and the morning were the second day. 9 And God said, Let the waters under the heaven be gathered together unto one place, and let the dry land appear: and it was so. 10 And God called the dry land Earth; and the gathering together of the waters called the Seas: and God saw that it was good. 11 …12 …13… And the evening and the morning were the third day. 14 And God said, Let there be lights in the firmament of the heaven to divide the day from the night; and let them be for signs, and for seasons, and for days, and years: 15 And let them be for lights in the firmament of the heaven to give light upon the earth: and it was so. 16 And God made two great lights; the greater light [Sun] to rule the day, and the lesser light [Moon] to rule the night: he made the stars also. 17 And God set them in the firmament of the heaven to give light upon the earth, 18 And to rule over the day and over the night, and to divide the light from the darkness: and God saw that it was good. 19 And the evening and the morning were the fourth day" (King James Version).

With these findings, I also established that the Biblical days of creation were literally 24 hours each. Because, in the 3500-year-old Biblical account of creation, Moses repeatedly said that Elohim (the God of Israel) created the universe, I deduced that the match between the scientific evidence of the formation of the universe and the Biblical account of creation attests that the God of the Bible is the creator, just as Moses testified. Considering all the variables involved in the split-gathering of the celestial bodies, it appeared to me, and I provided more proofs in other books, that if the initial events at the origin of the universe were not precisely calibrated, a different universe could have been formed. In other words, a different tuning at the beginning of the universe could have led to a different distribution and characteristics of bodies in the universe. This means that the turbulence and all the processes that led to the formation of the Solar System were calibrated to support life on Earth. In another book, I expounded on the origin of life. Therefore, at this time, I would like to thank God for creating such a beautiful world and for giving me the grace and wisdom to spend about 9 years decoding and writing books about the beginning.

In closing, I would like to say that one of the problems people have had in estimating the universe's age in billions of years is that they did not know how the world was formed. Instead of embracing the processes of the formation of the universe systematically, involving many processes at the same time, people took the short and wrong road by viewing natural processes linearly, meaning as if one

Nathanael-Israel Israel: Acclaimed at the Standout Scientific Authority Who Accurately Decoded the Universe-Origin

"simple" thing led to another one with a higher complexity, and so on and so forth, as if that was how complexity arose in nature. This way of thinking prevented many people from seeing reality from the mother of all turmoil's perspective, which is key to unlocking the code of this universe's formation. Unfortunately, because some people set their minds on the wrong things, even if every piece of proof is given to them about the universe, they will never believe or accept the facts. For what "moves" some people is not the truth, but lies and other things that they want, even if those things will eventually destroy them. Hence, in the name of "feeling good," some people have sold their lives in exchange for precarious pleasures, which reinforce their inability to properly use their brains to make decisions that will benefit them in the long term. Therefore, I urge you to revisit your belief or philosophy concerning the origin of the universe, for sooner or later it will affect your life in this world and in that to come, a fact some people deny while others have embraced. The universe had a beginning and a Creator, who set rules that you need to follow, or else your eternal future may be very bad! I know some people do not want to hear such a message, but I cannot conclude this book without warning you about the consequences of denying the existence of a Creator, as I have demonstrated in this book and in others I have written on the origin of the universe and everything in it.

Decrypting and explaining the mysteries of the universe's origin and formation is not a simple task that can be fully addressed in a single book. For if every detail of how the universe was formed and how it has been functioning could be written in books, I do not think the whole universe has enough space to contain them. Therefore, what I presented in this book is a quick glimpse of what happened (which I think can be more holistically apprehended through faith rather than relying solely on mathematical modeling, which lacks the proper languages, resources, and terms to decode the real story of the beginning). I hope this book "answers" some of the questions you may have about how the universe was formed.

I know that some people have a lot of questions concerning the origin of life and the organization of living things. However, due to the scope of this book and the number of details needed to properly explain life, I reserved a completely different book for that. I also understand that some people are eager to learn about the age of the universe and the fate and end of the world. Although I extensively addressed those subjects in other books, I did not talk about them much here, for not only would I not be able to clearly explain my thoughts regarding that topic using the language I assigned for this book, but also I also, I preferred sticking to my objective of this book, which is to appeal to the general public, hoping that those who are interested in the high-level scientific aspects or the Biblical aspects can follow up with the books I wrote on them: "*Turbulent Origin of the Universe*" and "*Reconciling Science and Creation Accurately,*"" respectively. Just as the data surrounding the formation of the universe is said to be very complex, so also are human beings very complex, and if I did not break down my books according to the target of each, I could have well found something worth sharing with the world, but the language used may have prevented most people from finding and dealing with the version of my books that interests them.

In addition to *"Turbulent Origin of the Universe*," *"Turbulent Origin of Chemical Particles*," and *"Turbulent Origin of Life*," I wrote another book ("Origin of the Spiritual World") on the hidden secrets (lost and rejected) about the origin of the universe, which were known thousands of years ago but which were not and are still not widely available to the general public. In that book, I used information considered as lost or rejected, known to a few people thousands of years ago, but to which less attention has been paid, at least from the scientific perspective. I was astonished that much of the information in the hidden books concurs with advanced scientific data discovered thousands of years after their revelation by ancient people who lived on this Earth long before anyone could ever think that anything called "science" could ever be invented to challenge them. Because many people who will be reading this book are not "religious" and even many of those who are religious do not believe in the pseudepigrapha (lost and rejected scriptures), I did not dwell on them in this book, but I felt obligated to point to them here so that those who would like to know more could identify where to look. Remember to consult the website (www.Science180.com) associated with my books, where I will provide updates on my work as well as additional information not in this book. To make a long story short, the current book is like a snapshot or a brief synopsis of the detailed demonstrations I did in my other books, and to get a comprehensive understanding of the beginning, you may need to consult those books as well. I hope this book has blessed you and improved your understanding of the beginning.

To you who have read this book from the beginning to this point, I want to thank you and congratulate you on reaching this milestone in your search for the truth about the origin of the universe. Because you read this book, I know your life and understanding of the origin of the universe will never be the same again. You are blessed, and therefore, walk in this blessing and boldly share the truth with others for the glory of God, to whom belongs the glory forever and ever.

'Science180 Academy' Success Strategy:
SCIENCE180 MODELS OF THE ORIGIN OF THE UNIVERSE AND ITS CONTENT

Science180 Models consist of all the theories elaborated by Nathanael-Israel Israel regarding his ground breaking discovery on the origin of the universe and its content including all forms of life and chemical particles. These theories are detailed in various books written by Dr. Nathanael-Israel Israel encompass the following:

1. *SCIENCE180 MODEL OF COSMOLOGY*, also called Science180 Cosmology, Science180 Model of Cosmology, Science180 Cosmological Model, a scientific theory that explains Science180 to the scientists. Discover the details of this model in Nathanael-Israel Israel's book titled *"Turbulent Origin of the Universe"*. In that book, you will also unearth the new physics that will revolutionize science forever and land you into a zone of original ideas that improve lives nonstop regardless of your expertise.

2. *SCIENCE180 CREATIONISM*, also called Science180 Model of the Creation of the Universe and Life by God, a scientific theory that presents the origin of the universe in a biblical language. If you want to learn more about how to scientifically prove the Biblical account of the creation of the universe and the existence of God in a way that makes the head of God deniers to spin faster than a DJ's turntable, then get Nathanael-Israel Israel's book titled *"Reconciling Science and Creation Accurately"*.

3. *SCIENCE180 MODEL OF THE ORIGIN OF CHEMICAL PARTICLES*, a scientific theory that explains the origin of chemical particles with the perspective of Science180 Turbulence. If you want to professionally learn how to transform the true knowledge of the origin of chemical particles into insights that significantly add value to your life in less time, successfully establish you as a symbol of freedom, power, creativity, and originality in your field of expertise, get Nathanael-Israel Israel's book *"Turbulent Origin of Chemical Particles"*, THE ultimate how-to guide for great people wanting to correctly decode the origin of the chemicals and positively transform their lives. Get this celebrated book today. Don't wait!

4. *SCIENCE180 MODEL FOR THE GENERAL PUBLIC* (which explains the origin of the universe and life to the general public in a language that laypeople can understand). Find out more in Nathanael-Israel Israel's book called *"From Science to Bible's Conclusions"*, a scientifically verifiable, bestselling book to finally get the accurate, jaw-dropping answer that has been rationally shaking believers, skeptics, and freethinkers. Get this very popular book today.

5. *SCIENCE180 MODEL OF LIFE-ORIGIN*, or Science180 Model of the Origin of Life, a scientific theory that explains the origin of all forms of life using turbulence. To unlock the step-by-step pathway to decode the origin of life and get the power, freedom, and boldness to detect, correct, and remove all misinformation, ambiguity, and misleading claims and theories surrounding the life-origin and take advantage of the opportunities that an accurate understanding of the life-origin creates, get Nathanael-Israel Israel's book titled *"Turbulent Origin of Life"*.

6. *SCIENCE180 MODEL FOR CHILDREN*, a children's version of the theory of the origin of the universe and life in a language that 7-12 years old children can properly understand. To know the proven formula that helps children to easily answer their huge universe-origin and life-origin questions with confidence, humor, and joy, get *"How Baby Universe Was Born"*, the pragmatic book that has been causing children to belly laugh and thank those who offered it to them.

7. *SCIENCE180 MODEL OF PSEUDEPIGRAPHA*, a deep explanation of the secrets of the origin of the universe and life revealed a long time ago, but hidden from the general public. To discover how the only one ancient blueprint has the reliable power to help you to accurately decrypt the spiritual origin and history of everything in the universe, get Nathanael-Israel Israel's book called *"Origin of the Spiritual World"*. In it, you will discover deep rejected secrets that have prevented humankind from unearthing the beginning of the universe and know how to properly use the lost and rejected scriptures to articulate the process by which the universe was formed, so you can use that insight to improve your understanding of the Bible, innovate in your domain of interest, and improve your life perpetually.

Nathanael-Israel Israel: Acclaimed at the Standout Scientific Authority Who Accurately Decoded the Universe-Origin

8. *SCIENCE180 MODEL OF THE PROOF OF THE EXISTENCE OF GOD,* a theory that ties together most of Nathanael-Israel Israel's discoveries that scientifically prove the existence of God. With Nathanael-Israel Israel's book *"Science180 Accurate Scientific Proof of God"*, you will surely know the only way to scientifically know if God exist, and if so, which of the thousands of beings worshipped across the globe is the true God. In that book, you will also discover the errors in the scientific and religious theories (about the origin of the universe, life, and chemicals) that are putting you at a high risk you will never recover from if you don't quickly and confidently learn how to rationally take control over threats lurking at the edge of your efforts to understand the universe and life today.

9. SCIENCE180 THEORY OF EVERYTHING, (also called the theory of all theories), ties together everything in the universe into a single theory. Checkout Science180.com to learn more about the incoming book that covers this extremely important topic.

Nathanael-Israel Israel: Acclaimed at the Standout Scientific Authority Who
Accurately Decoded the Universe-Origin

NEXT STEPS OF THE JOURNEY

Get free resources on Science180.com

If you have finished reading this book and would like to learn more about my discoveries and how they can help you, you are at the right place. Indeed, I am really committed to helping you address any questions you may still have about the origin, function, and fate of the universe, and how you can partner with me to achieve greater results.

To get free resources that will help you understand other aspects of the universe formation not covered in this book, visit Science180.com and my personal website, Israel120.com. On those sites, I will be sharing guides and strategies to get the most out of my initiatives. I will also be sharing my favorite references, tips, next steps, readings, and other important things in the pipeline that will help you, regardless of your field of expertise, interests, or needs.

Subscribe to "Science180 Newsletter": The only accurate universe-origin, life-origin, and chemicals-origin newsletter in the whole world!

Be a part of decoding the universe-origin, life-origin, and chemicals-origin! Get origin-related news, information, discoveries, updates, announcements, reviews, articles, educational materials, and opportunities, from a holistic perspective not available anywhere else, so you can participate in and enjoy decoding the origin, current state, and fate of the universe and its content. You will also receive priceless tips about how Nathanael-Israel thinks, what his secrets and initiatives are, what he has accomplished, and what he recommends. Without any delay, sign up for the Science180 Newsletter today at Science180.com/newsletter. It is free!

Speaking engagement

In addition to writing groundbreaking books and engaging in other business endeavors, Nathanael-Israel Israel is a renowned speaker whom you can invite to speak at your organization.

Values that Dr. Nathanael-Israel Israel can add to your life include the following:

- Rare expertise and tips that will increase your abilities
- Usefulness that will advance your impact regardless of your field of expertise
- Understanding of the world that will sharpen your perspective
- Critical information that will positively change your life
- Experiences turned into insight that will motivate and guide you
- Irrefutable scientific proofs of the existence of God that will save you time and launch you into a zone of unlimited opportunities
- Unquestionable scientific proofs of how God created the universe
- Accurate demonstration of the historic formula that reconciled science and the Bible
- Enlightenment that will help people to start using their brains instead of just praying and expecting God to do everything for them

For speaking inquiries, including how to book Dr. Nathanael-Israel Israel to speak to your organization or at an event, visit Science180.com/speaking for more details.

As the standout scientific authority who accurately decoded the universe, Nathanael-Israel Israel has been helping countless people across the globe discover and understand the complex origin of the universe without overlooking the challenging questions that people of all ages have been struggling to answer for thousands of years! As the true go-to expert on the formation of the universe and life, Nathanael-Israel believes that, regardless of age, background, culture, religion, or profession, everyone deserves to understand how the universe and life were formed and how they can leverage that knowledge to improve lives nonstop. Therefore, his groundbreaking discoveries of the formation of the universe, life, and chemicals have been broken down into books tailored to scientists (including physicists, chemists, biologists, mathematicians), laypeople, or the general public, general internationally acclaimed thinkers, philosophers, children, etc., therefore maximizing the benefits to humanity. These historic, internationally-acclaimed origin books include:

- "Turbulent Origin of the Universe"
- "Reconciling Science and Creation Accurately"
- "Turbulent Origin of Chemical Particles"
- "From Science to Bible's Conclusions"
- "Turbulent Origin of Life"
- "Origin of the Spiritual World"
- "How Baby Universe Was Born"
- "How God Created Baby Universe"
- "Science180 Accurate Scientific Proof of God"

When you hire Nathanael-Israel Israel to speak at your organization, you will:

- Get specific in-depth knowledge, up-to-the-minute information, ideas, and insights about the universe's origin, life's origin, and chemicals' origin so that you expand your market, cut useless costs, stop wasting time on inadequate projects, and start focusing on the profitable solutions
- Get relevant universe-origin stories that are specific to your field of expertise
- Learn from a cooperative, flexible, and easy-to-work-with expert who will respond to your universe formation needs and position you to stay on top of your competitors
- Interact with a renowned expert who will not just lecture you, but will help you sort out your origin-related questions using strategies to tap into deep secrets you ignore
- Listen to an experienced expert who discovered outstanding secrets about the origin of all there is
- Learn authentic information not from someone who just reads you a PowerPoint, but from the true go-to expert (when it comes to critical cosmological problems) who will share with you both his mistakes and successes that will help you get much closer to the better life you want to live
- Revolutionize every origin-related domain with your accurate understanding of the universe's origin.
- Scientifically learn how the Earth was formed on the 3rd day of creation
- Logically, learn how the Sun and the Moon were formed on the 4th day of creation
- Hear Dr. Nathanael-Israel Israel's personal selection and teaching of key topics that will help you break the code of the universe's formation and functioning; strategically enlighten you; guide you to navigate and filter the massive data collected on the universe and its content so you know how to answer the world's most challenging origin questions; remove any scientific and philosophical cataracts that may be blocking you; and help bring you many steps closer to your best life today and forever
- Hear the greatest scientific and philosophic lessons of some top scientists, philosophers, thinkers, and public figures who have realized historic mistakes they made in life (concerning the origin of the universe, life, and chemicals), and that they corrected thanks to the discoveries of Nathanael-Israel Israel, who founded Science180, and who is acknowledged as the scientist that truly decrypted the universe-origin for the first time
- Get world-key lessons successful people have learned in life, and how people can learn from their experiences to improve their lives instead of repeating their mistakes that many people still ignore at their own peril.

To book Dr. Nathanael-Israel Israel for a speaking engagement, visit Science180.com/speaking.

How you can make money by joining the affiliate program to sell Nathanael-Israel Israel's books

Greetings,

Do you want to make easy money by selling the #1 universe-origin, life-origin, and chemicals-origin books on your website, newsletter, and by mail? You can start making some money by helping sell Science180 Books, including this one, on your website and through your network. Indeed, by now you know that I operate a website called Science180.com, which specializes in helping people around the globe scientifically decode and understand the formation of the universe, life, and chemicals.

Your contacts, site, blog, forum, podcast, and newsletter may be admired among my target audience. Some of my products and services may be of interest to your audience. My books are the first in history to scientifically demonstrate the match between science and Biblical creation in a way that satisfies both believers and nonbelievers, a historic achievement and discovery that is revolutionizing our view of the origin of the universe, life, and chemicals for the benefit of humankind.

Imagine you have a website where you can talk to people about my books and services, and get a great percentage of every purchase they make on my site? Imagine you send a link to my books to your friends or network, and when any of your contacts buy a copy, you get a percentage of what they pay on my sites. Imagine you can email your friends to spread the good news about my books, and when anyone uses that link to buy them, I give you something. Well! This is what the affiliate program is about. Apply today or learn more about it at Science180.com/affiliate. Likewise, if you own a website, you can apply for Science180's affiliate program, and I will send you a specific affiliate link that you will place on your website and newsletter, and if people click on it to buy my books, they will be led to my page, and after they buy, I will pay you a certain amount, sharing the profit with you instead of just verbally saying thank you.

Would you be interested in reviewing some of my products and services to explore becoming an affiliate? We have a wonderful affiliate program, and commissions are paid quickly and accurately.

If you are satisfied with the quality of my products and services, I am convinced you will also be impressed by our affiliate program.

I look forward to hearing from you

Nathanael-Israel Israel, PhD

Collaborate or partner with Nathanael-Israel Israel

If you have any lawful idea, initiative, or suggestion for a genuine partnership with Dr. Nathanael-Israel Israel or Science180, please visit Science180.com/partner to inform us.

How to be trained or mentored by or have a one-on-one consultation

with Dr. Nathanael-Israel Israel

Hire Nathanael-Israel Israel to train you or your organization in the best ways to conduct yourself and your organization, and to align your organization with the real understanding of the origin of the universe, of life, and of chemical particles, in a way you will not hear anywhere else. Nathanael-Israel Israel offers training through the "Science180 Academy" program. For training purposes, please visit Science180Academy.com.

Visit Nathanael-Israel Israel's personal website to get great resources for free that you won't find anywhere else

To stay in touch with Dr. Nathanael-Israel Israel and get updates directly from him, please visit his website, Israel120.com, and sign up for his popular newsletter at Israel120.com/newsletter for free.

Ask for a review

If you are a book reviewer or a professional wanting to review this book or others written by Nathanael-Israel Israel, please contact us at Science180.com/AskForReview

Donate and support Nathanael-Israel Israel's efforts and initiatives

To help humankind accurately understand the real origin of the universe and its content, as I have done in the groundbreaking books I published after 12 years of sacrifice, I need your financial support. Please consider donating to me by visiting Israel120.com/donate.

Your donation will help me continue doing what I did to bring these books to life, which you enjoyed and know will help many people across the globe. No amount of money is too small or too big. Whatever you can give, please give.

Quantity discounts: Purchase Science180 books including this one, in bulk at a special discount

To purchase Science180 books, including this one, in bulk at a special discount for sales promotion, corporate gifts, fund-raising, or educational purposes, or to create special editions to specifications, contact specialsales@science180.com or visit Science180.com/discount.

Buy a copy of Nathanael-Israel Israel's books for your friends, family, or someone else

If this book has been a blessing to you, and we know it has, please consider getting another copy and giving it to a friend, a family member, or someone you think it may help or challenge. If you want to get many copies, we can even give you a discount;

just contact us as we previously explained.

Recommend Nathanael-Israel Israel's books to your organization

Because I know this book has been a blessing to you, I ask that you recommend it, along with others I wrote, to your organization, class, workplace, church, school, network, or clubs. Recommending this book will help others to tap into the blessing and opportunities that my books will open for them.

Share Nathanael-Israel Israel's groundbreaking discovery with others

To improve more lives, please share the findings of Nathanael-Israel Israel's books with others, for many people out there still do not understand how the universe was formed, and sharing your experience of reading this book will help them. If you enjoy Nathanael-Israel Israel's books, please help other people find them by writing a book review on your blog or on online bookstores, or write one and share it with us. Likewise, share and mention this book on your social media platforms (e.g., Facebook, Twitter, YouTube, etc.).

Follow Nathanael-Israel Israel on social media

In our modern world, social media has become a major factor in how messages spread across the globe. To ensure more people hear about the good news revealed in my books, I need you to follow me and share my content on your social media and in your network. To know the full list of my social media accounts and follow me, please visit Science180.com/socialmedia.

Share your feedback, criticism, testimony, experience, adventures, story, or comment about this book with me

How have Nathanael-Israel Israel's books and services at Science180 improved your life? I would love to hear from you.

To better understand how I can help you next and encourage others, I need to capture your testimony or criticisms. Please visit the feedback page, Science180.com/feedback, to tell me:

- How this book impacted you or will impact you
- What you like or dislike or disagree with
- What do you think, wish, or dream that I need to work on next
- What you wish to see in this book, but that was absent
- What shocked you the most
- What got your heart pumping as you were reading this book
- What you found most insightful or thought-provoking
- What do you want to do to be a part of my journey
- How my work changed your life or someone else's life

Message from the publisher of this book

Just like Nathanael-Israel Israel, you can publish your book(s) with us, too. To get started and see how we may help you, please visit Science180Publishing.com today.

To contact Nathanael-Israel Israel or Science180

For any suggestions or questions, please visit Science180.com/contact and Nathanael-Israel Israel's personal website, Israel120.com. Feel free to ask me any questions you have about the formation of the universe, life, and chemistry.

Another Book by Nathanael-Israel Israel:
ORIGIN OF THE SPIRITUAL WORLD

ONLY ONE ANCIENT BLUEPRINT HAS THE RELIABLE POWER TO HELP YOU TO ACCURATELY DECRYPT THE SPIRITUAL ORIGIN AND HISTORY OF EVERYTHING IN THE UNIVERSE

Countless books talk about the origin of the universe and of life, but this amazing book is the first and the only one that has undeniably explained how the formation of the universe and everything in it was truly revealed in the rejected and hidden scriptures such as the Books of Enoch and others. In "*Origin of the Spiritual World*", you will:

- Discover deep rejected secrets that have prevented humankind from unearthing the beginning of the universe
- Plainly see the scientific proof (hidden in scriptures) of the formation of the Earth, the Moon, and the Sun in a matter of days, a historic revelation that bizarrely and shockingly matches the scientific data as scientifically proved in "*From Science to Bible's Conclusions*", a popular book written by Dr. Nathanael-Israel Israel
- Properly use the lost and rejected scriptures to articulate the process by which the universe was formed, and use that insight to improve your understanding of the Bible, innovate in your domain of interest, and improve your life perpetually
- Empower and align yourself with the historic breakthrough that has done what no other discovery has ever done: accurately unlock and decode mysteries concerning the origin of the cosmos and its content using scientific keys revealed in ancient scriptures that some elites have concealed (*Science180.com/pseudepigraphic*)
- Discover and apprehend the complex formation of the universe and life without leaving out the challenging questions that people of all ages have been struggling to answer for thousands of years, while the answers were hidden

- Find more joy in life through a clear interpretation of old and fresh revelations about the creation of the universe astonishingly backed by modern science, which some people wrongly think opposes the Bible
- Make a difference and blaze new trails for those who depend on your leadership

If you believe in God, have some origin-related questions which answers you cannot find anywhere, not even in the Bible, and if you want to tap into historically neglected revelations to answer fundamental universe and life questions, then be sure to get a copy of *"Origin of the Spiritual World"* today.

Dr. Nathanael-Israel Israel happens to be the discoverer of the historic mathematical equations that scientifically demonstrated that the Earth was formed 2.82 days, the Moon 3.32 days, and the Sun 3.69 days after the beginning of the universe, therefore confirming the Biblical account of creation that revealed about 3500 years ago that the formation of the Earth was completed on the 3rd day, while that of the Moon and the Sun was completed on the 4th day of creation. Nathanael-Israel Israel is referred to as the "Undisputable Specialist of all Questions at the Intersection of Science and Biblical Creation". Learn more about this rare scientist at Israel120.com.

'Science180 Academy' Success Strategy
HOW TO RAISE RATIONAL CHILDREN IN OUR MODERN WORLD

In our modern secular world, and with the many things that kids are taught at school and over which parents have little control once the kids head to public school, parents have a lot to worry about. But it does not have to be that way. Universe-origin and life-origin scientist Dr. Nathanael-Israel Israel has discovered that, more than ever, parents have a crucial responsibility to rationally prepare their kids to have a strong worldview that properly embraces both science and faith, so their kids are not pulled on one side by the secular education and on the other side by religious belief. But how can parents and their children achieve that common goal?

Listen to this Beninese-American scientist and mathematician Dr. Nathanael-Israel Israel to figure it out. Nathanael-Israel is the author of the acclaimed book *"How Baby Universe was Born"*, an easy to understand scientific book primarily written for children age 7-12 years old to help them properly crack the code of the formation of the universe in a language they completely enjoy, and that prepares them to fight any secular or religious theory that may try to rationally drift them away from the reality of everything!

Sample questions that will get answered include the following and many more:

- How can parents use the latest breakthrough about the universe-origin to rationally raise their kids?
- How can parents prepare their children from being victims of the danger of wrong theories and dogmas on the origin of life and the universe?
- What can parents do to shield their children from the influence of religious and scientific beliefs that try to enslave them in the name of reason or faith?
- Why is wrong science not the only danger of raising rational children, but wrong belief as well?
- How can we help children to positively navigate the intersection of science and faith?

Learn more at Science180.com/children

Nathanael-Israel Israel: Who Happens to be the World's #1 Authority on the
Turbulent Origin of the Universe and Life

REFERENCES

Eggers Jens and Emmanuel Villermaux. 2008. Physics of liquid jets. Rep. Prog. Phys., 71(3):036601 (79pp). doi:10.1088/0034-4885/71/3/036601.

George W. K. (2013). Lectures in Turbulence for the 21st Century. Department of Aeronautics Imperial College of London, London, UK, and Department of Applied Mechanics - Chalmers University of Technology Gothenburg, Sweden. Retrieved on October 19, 2017, from www.turbulence-online.com.

Israel Nathanael-Israel (2025a). Turbulent Origin of the Universe. Science180, Augusta, USA 683 pages.

Israel Nathanael-Israel (2025b). From Science to Bible's Conclusions. Science180, Augusta, USA 170 pages.

Israel Nathanael-Israel (2025c). Reconciling Science and Creation Accurately. Science180, Augusta, USA 299 pages.

Israel Nathanael-Israel (2025d). Turbulent Origin of Chemical Particles. Science180, Augusta, USA 397 pages.

Israel Nathanael-Israel (2025e). Turbulent Origin of Life. Science180, Augusta, USA 370 pages.

Israel Nathanael-Israel (2025f). Origin of the Spiritual World. Science180, Augusta, USA 151 pages.

Israel Nathanael-Israel (2025g). How Baby Universe Was Born. Science180, Augusta, USA 130 pages.

Israel Nathanael-Israel (2025h). How God Created Baby Universe. Science180, Augusta, USA 224 pages.

Israel Nathanael-Israel (2025i). Science180 Accurate Scientific Proof of God. Science180, Augusta, USA 214 pages.

Israel Nathanael-Israel (2026a). Mathematical Proof of God's Existence at the Intersection of Science and Faith. Science180, Augusta, USA 74 pages.

Malvern (2016). Malvern Instruments White Paper - A Basic Introduction to Rheology, www.malvern.com, Worcestershire, UK, 20 pages.

NASA (2018). Planetary fact sheets. Fact sheets of the Sun, planets, satellites, rings, and selected asteroids in the Solar System. Author/Curator: Dr. David R. Williams,

NASA Goddard Space Flight Center, Greenbelt, MD, USA. Retrieved on November 19, 2018, from http://nssdc.gsfc.nasa.gov/planetary/factsheet/.

Petitjeans Philippe and Frédéric Bottausci (2020). Structures tourbillonnaires étirées : les filaments de vorticité. Laboratoire de Physique et de Mécanique des Milieux Hétérogènes (UMR CNRS 7636) Ecole Supérieure de Physique et de Chimie Industrielles 10, rue Vauquelin, 75005 Paris. 13 pages.

Price J. F. (2006). Lagrangian and Eulerian Representations of Fluid Flow: Kinematics and the Equations of Motion. Woods Hole Oceanographic Institution, Woods Hole, MA, 02543. http://www.whoi.edu/science/PO/people/jprice June 7, 2006.

Yinon Bentor (2016). Chemical Elements.com. Retrieved on February 27, 2023, from http://www.chemicalelements.com.

INDEX

Properly Understand Turbulence. Only on Science180

Nathanael-Israel Israel: Universe-origin Doctor / Expert, Consultant, Speaker,
Scientist, Author

Nathanael-Israel Israel: Universe-origin Doctor / Expert, Consultant, Speaker,
Scientist, Author

Nathanael-Israel Israel: Universe-origin Doctor / Expert, Consultant, Speaker, Scientist, Author

ABOUT THE AUTHOR

Dr. Nathanael-Israel Israel scientifically challenged, demystified, and changed the way scientists, nonscientists or laypeople, believers, unbelievers, and all kinds of freethinkers (e.g., atheists, humanists, skeptics, evolutionists, and anti-creationists) think about the formation of the universe, life, and chemicals. Using a unique insight that courageously challenges conventional scientific methods and religious boundaries, Dr. Israel has earned the reputation as the world's #1 authority on complex issues pertaining to the origin of the universe, life, and chemical particles. Because no one has ever imagined that a human being could accurately decipher the blueprint for the formation of the universe as Nathanael-Israel Israel did, people of all scientific, religious, freethinking, and political backgrounds have been seeking his historic expertise to decode the origin of the universe in unconventional ways and improve lives nonstop.

Dr. Nathanael-Israel Israel is a member of the American Chemical Society, American Association for the Advancement of Science, American Society of Agricultural and Biological Engineers, American Society for Microbiology, American Society of Biochemistry and Molecular Biology, Ecological Society of America, American Society of Agronomy, Crop Science Society of America, and Soil Science Society of America. A renowned personality in the universe-origin, life-origin, and chemicals-origin space, Dr. Nathanael-Israel Israel is the founder of Science180, the American organization that operates Science180 Academy (Science180Academy.com), a non-degree training, speaking, consulting, and mentoring program designed to groom and empower people of all backgrounds in the truth about the origins of the universe, life, and chemicals. Before launching Science180, Dr. Nathanael-Israel Israel worked as a scientist at a major Fortune 500 biotechnology company in the USA, where this Beninese-American also founded and owns a news company. Some of his groundbreaking books include:

- Turbulent Origin of the Universe
- Reconciling Science and Creation Accurately
- Turbulent Origin of Chemical Particles

Nathanael-Israel Israel: Who happens to be the World's #1 Authority on the Turbulent Origin of the Universe and Life

- Origin of the Spiritual World
- From Science to Bible's Conclusions
- Turbulent Origin of Life
- How Baby Universe Was Born
- How God Created Baby Universe
- Science180 Accurate Scientific Proof of God
- Mathematical Proof of God's Existence at the Intersection of Science and Faith.

Get these thoughtful books to figure out what happened at the beginning of the universe, what is coming up, and why it is time to urgently rethink everything you have been told about the origin of the universe, so you don't eventually regret it! Connect with this historic scientist and get free resources today by visiting Israel120.com.

Nathanael-Israel Israel: Historic Discoverer of the Formula to Accurately Decode the Origin of the Universe, of Life, and of Chemicals in a Few Days

PLEASE FEEL FREE TO WRITE YOUR QUESTIONS HERE. I WILL LOVE TO HEAR FROM YOU ABOUT THEM. PLEASE CONTACT ME AT WWW.ISRAEL120.COM/CONTACT

Nathanael-Israel Israel: Who happens to be the World's #1 Authority on the Turbulent Origin of the Universe and Life

Nathanael-Israel Israel: Historic Discoverer of the Formula to Accurately
Decode the Origin of the Universe, of Life, and of Chemicals in a Few Days

PLEASE FEEL FREE TO WRITE YOUR QUESTIONS HERE. I WILL LOVE TO HEAR FROM YOU ABOUT THEM. PLEASE CONTACT ME AT WWW.ISRAEL120.COM/CONTACT

Nathanael-Israel Israel: Who happens to be the World's #1 Authority on the Turbulent Origin of the Universe and Life

Nathanael-Israel Israel: Historic Discoverer of the Formula to Accurately
Decode the Origin of the Universe, of Life, and of Chemicals in a Few Days

PLEASE FEEL FREE TO WRITE YOUR QUESTIONS HERE. I WILL LOVE TO HEAR FROM YOU ABOUT THEM. PLEASE CONTACT ME AT WWW.ISRAEL120.COM/CONTACT

Nathanael-Israel Israel: Who happens to be the World's #1 Authority on the Turbulent Origin of the Universe and Life

Nathanael-Israel Israel: Historic Discoverer of the Formula to Accurately
Decode the Origin of the Universe, of Life, and of Chemicals in a Few Days

PLEASE FEEL FREE TO WRITE YOUR QUESTIONS HERE. I WILL LOVE TO HEAR FROM YOU ABOUT THEM. PLEASE CONTACT ME AT WWW.ISRAEL120.COM/CONTACT

Nathanael-Israel Israel: Historic Discoverer of the Formula to Accurately
Decode the Origin of the Universe, of Life, and of Chemicals in a Few Days

www.ingramcontent.com/pod-product-compliance
Lightning Source LLC
Chambersburg PA
CBHW070921130626
46555CB00001B/236